AF718532

Numerical Analysis
Demonstrations on the HP-33E

Peter Henrici

Marie Louise Henrici

E. W. GIRARD

1807 175 YEARS OF PUBLISHING 1982
John Wiley & Sons
New York Chichester Brisbane Toronto Singapore

Copyright © 1982 by John Wiley & Sons, Inc.

All Rights Reserved.

Reproduction or translation of any part of this work beyond that permitted by Section 107 or 108 of the 1976 United States Copyright Act without the permission of the copyright owner is unlawful. Requests for permission or further information should be addressed to the Permissions Department, John Wiley & Sons, Inc.

ISBN 0 471 05943 9
Printed in the United States of America

10 9 8 7 6 5 4 3 2 1

PREFACE

This brochure is conceived as a supplement to the textbook by the senior author, ESSENTIALS OF NUMERICAL ANALYSIS, WITH POCKET CALCULATOR DEMONSTRATIONS. It contains HP-33E programs for virtually all demonstrations given in that text, and for most of the problems requiring numerical work. The documentation of the demonstration programs is sufficiently self-contained so that they can be used without previous programming experience. The presentation of the programs for the solution of homework problems is somewhat more concise; here some additional effort may be required for the complete understanding of the program.

Theory in numerical analysis is never (or should never be) an end in itself. Its ultimate goal is the creation of efficient and stable algorithms for the constructive solution of mathematical problems. Conventional mathematical notation usually provides only a very incomplete description of an algorithm. It does not account, for instance, for the storage requirements of an algorithm, or for the precise order in which the arithmetical operations are to be performed. To understand an algorithm completely, one has to program it. Working within the limitations of a pocket calculator provides the additional challenge that the programmer must concentrate on the essential features of an algorithm, and that he must pay particular attention to the allocation of storage. In this sense, a study of the programs in this collection should help the student to acquire the habit of writing efficient and lean programs for scientfic computation.

Even if an algorithm has been programmed with perfect logic and economy, it is by no means certain that it will perform satisfactorily in practice. Although theoretical numerical analysis will offer some guidance, such essential features as the efficiency, the reliability, and the numerical stability of an algorithm can be judged with finality only on the basis of tests in concrete situations, including situations with unusual or extreme sets of data. One of the benefits for the users of these programs consists in the immediacy and concreteness in which they will experience the performance of the algorithms presented.

Many algorithms in this collection will, within the obvious physical limitation of the pocket calculator, produce satisfactory answers to numerical problems. In other algorithms, the data sets are too small to be useful in realistic situations. These algorithms then may at least serve as models for how to solve such problems on larger computers. Other programs are designed to illustrate the typical pitfalls of numerical computation such as cancellation, smearing, and numerical instability.

Regrettably, some areas of numerical analysis, such as numerical linear algebra and the numerical treatment of partial differential equations, cannot as yet be realistically illustrated on pocket calculators because of the large sets of data that would have to be processed. The presentation of such problems in the format of this book will have to wait until more powerful calculators are available.

We wish to express our thanks to a number of individuals who have contributed ideas to the programs presented here. Dr. O. Pretzel (London) has given permission to include his ingenious Runge-Kutta program. W. Frangen (Karlsruhe) through his numerous published programs has expanded our vision of what can be done on pocket calculators. The staff of John Wiley & Sons, Inc., have responded most willingly to the particular requirements of this book. Once again, our thanks go to Brigitte Knecht for her unsurpassed ability to produce a camera-ready manuscript out of our scribblings.

Zurich, March 1981

P. Henrici
M.-L. Henrici

INTRODUCTION

Description of programs belonging to Demonstrations are arranged according to the following scheme:

1. Purpose
2. Method
3. Flow Diagram
4. Storage and Program
5. Operating Instructions
6. Examples

The statement of purpose usually is very brief. It informs the reader whether the program is intended for practical use, or whether it merely serves a didactic purpose.

The description of the method is strictly confined to what is being done. For analytical justification and for technical details, the reader is usually referred to the book ESSENTIALS OF NUMERICAL ANALYSIS, WITH POCKET CALCULATOR DEMONSTRATIONS, by P. Henrici, which is quoted simply as "Essentials". One reference is to ACCA I, which means vol. I of "Applied and Computational Complex Analysis" by the same author.

The flow diagrams do not follow any particular standard format. The intent is simply to make the structure of a program visible at a glance. In the case of some very simple programs the flow diagram is omitted.

The section entitled "Storage and Program", naturally, is the core of each program description. We use the symbol R_k to denote the storage register number k. Quantities that must be stored in the storage registers ahead of the computation are encased, thus:

x	y	z

The program listing uses the key symbols as they are found on the keys of the HP-33E calculator, with the exception of multiplication, which is indicated by * in order to avoid confusion with the letter "x". The listing is as compact as possible. For yellow and blue instructions it is understood that the keys "f" and "g" are to be pressed in advance; no ambiguity can arise by our abbreviated notation. The arrows → in front of certain instructions are not part of the program. They simply indicate an entrance from a GTO instruction.

At the expense of a certain amount of repetition, we have attempted to keep the operating instructions reasonably self-contained. We also give indications and explanations of possible failures of some programs.

For the examples we usually give references to the appropriate demonstrations in "Essentials". In a few instances, additional examples are provided.

The descriptions of programs for numerical problems usually consist only of the one part above entitled "Storage and Program". Additional explanations are kept to a minimum.

And now: Have fun!

Demonstration 1.2-2:

NUMERICAL COMPUTATION OF A SUM

1. Purpose

To show that on the computer simple laws of algebra, which in theory are valid for the real numbers, may not hold.

2. Method

By computing numerically the specific sums

$$s_n := 1 + \sum_{k=1}^{n} \frac{1}{k^2 + k}$$

(I) as usual, from the beginning and (II) from the tail end, we do not obtain the same numerical values for a fixed n . Hence the associative law in the additive group of real numbers is violated. See Essentials §1.2.

3. Flow Diagram

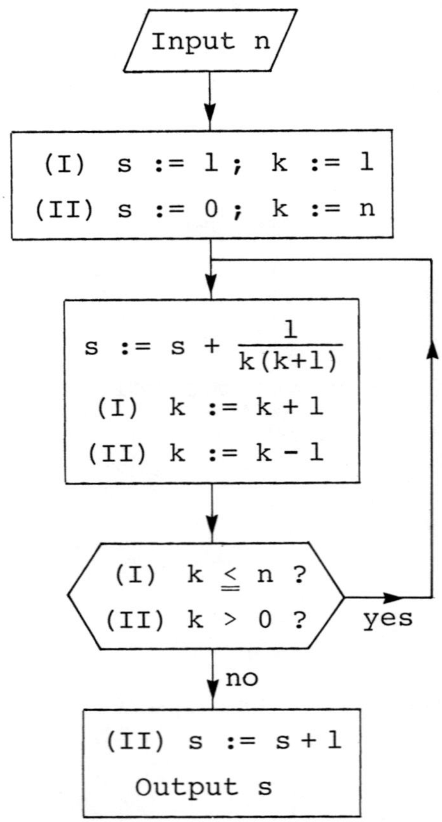

4. Storage and Program

(I)

R_0	R_1	R_2	R_3	R_4	R_5	R_6	R_7
k	s	[n]					

	00	10	20	30	40
0	☒	1			
1	1	STO+0			
2	STO 1	RCL 2			
3	STO 0	RCL 0			
4	→ RCL 0	$x \leq y$			
5	1	GTO 04			
6	+	RCL 1			
7	*	GTO 00			
8	1/x				
9	STO+1				

(II)

R_0	R_1	R_2	R_3	R_4	R_5	R_6	R_7
k	s	[n]					

	00	10	20	30	40
0	☒	STO+1			
1	CLX	1			
2	STO 1	STO-0			
3	RCL 2	RCL 0			
4	STO 0	$x > 0$			
5	→ RCL 0	GTO 05			
6	1	1			
7	+	STO+1			
8	*	RCL 1			
9	1/x	GTO 00			

5. Operating Instructions

Load program, move to RUN and load n into R_2. Upon pressing

FIX 9

PRGM

R/S

s_n will be computed and displayed. To obtain another s_n, load new value of n into R_2 and press R/S.

6. Example

See Essentials, Demonstration 1.2-2.

Demonstration 1.3-3:

NUMERICAL EVALUATION OF THE LENGTH OF A POLYGON

1. Purpose

To show that subtraction of two nearly equal numbers causes cancellation, see Essentials §1.3.

2. Method

We compute the length of the straight line segment joining the points (1,0) and (0,1) as special case of a formula for the length s_n of a polygon with vertices (r_k,ϕ_k), $k = 1, 2, \ldots, n$, i.e. by

$$\text{(I)} \qquad s_n = \sum_{k=1}^{n} \sqrt{r_k^2 - 2r_k r_{k-1}\cos(\phi_k-\phi_{k-1}) + r_{k-1}^2}$$

$$\text{(II)} \qquad s_n = \sum_{k=1}^{n} \sqrt{(r_k-r_{k-1})^2 + 4r_k r_{k-1}\sin^2\left(\frac{\phi_k-\phi_{k-1}}{2}\right)} \;,$$

see Essentials, Demonstration 1.3-3.

3. Flow Diagram

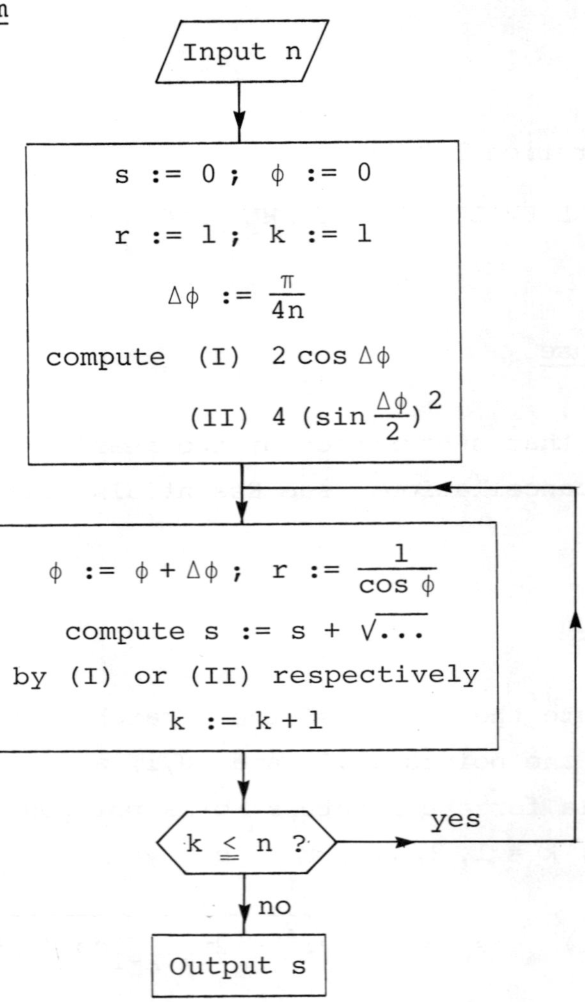

4. Storage and Program

(I)

R_0	R_1	R_2	R_3	R_4	R_5	R_6	R_7
n	r_{k-1}	r_k	ϕ_k	$\Delta\phi$	$2\cos\Delta\phi$	s	k

	00	10	20	30	40
0		RCL 0	GSB 43	RCL 5	GTO 18
1	RAD	*	RCL 1	*	RCL 6
2	CLX	÷	x^2	-	R/S
3	STO 3	STO 4	RCL 2	$\sqrt{x}$	RCL 3
4	STO 6	cos	x^2	STO+6	cos
5	1	2	+	1	1/x
6	STO 1	*	RCL 1	STO+7	STO 2
7	STO 7	STO 5	RCL 2	RCL 0	RTN
8	π	→ RCL 4	STO 1	RCL 7	
9	4	STO+3	*	$x \leq y$	

(II)

R_0	R_1	R_2	R_3	R_4	R_5	R_6	R_7
n	r_{k-1}	r_k	ϕ_k	$\Delta\phi$	$4(\sin\frac{\Delta\phi}{2})^2$	s	k

	00	10	20	30	40
0		RCL 0	STO 5	RCL 2	RCL 7
1	RAD	*	RCL 4	STO 1	$x \leq y$
2	CLX	÷	STO+3	-	GTO 21
3	STO 3	STO 4	GSB 45	x^2	RCL 6
4	STO 6	2	RCL 1	+	R/S
5	1	÷	RCL 2	$\sqrt{x}$	RCL 3
6	STO 1	sin	*	STO+6	cos
7	STO 7	x^2	RCL 5	1	1/x
8	π	4	*	STO+7	STO 2
9	4	*	RCL 1	RCL 0	RTN

5. Operating Instructions

Load program, move to RUN and load n into R_0. Upon pressing

FIX 9
PRGM
R/S

s_n will be computed and displayed. To obtain another s_n, load new number n of vertices into R_0 and press

PRGM
R/S

By rewriting subroutines 43 in (I) and 45 in (II) accordingly both programs can be used for other functions $r = r(\phi)$. If problems of programming space arise, instruction 01 and the assigning of initial values 02 ÷ 07 can be executed in the RUN mode before pressing PRGM for the first time.

6. Example

See Essentials, Demonstration 1.3-3.

Demonstration 1.3-4:

ARCHIMEDEAN DETERMINATION OF π

1. Purpose

To demonstrate the fact that one half of the length of the circumference of a regular n-gon inscribed into the unit circle as $n \to \infty$ tends to $\pi = 3.141592654...$. We impose the working rule that the only nonrational function which we are permitted to evaluate is the square root.

2. Method

Using elementary trigonometry and algebra one obtains the three following formulas for the determination of π; in each case $y_1 = 2$, $y_2 = 2\sqrt{2}$, for $k = 2, 3, ...$

$$y_{k+1} = 2^{k+1} \sqrt{\frac{1}{2}(1 - \sqrt{1-[2^{-k}y_k]^2}} \quad , \qquad \text{(I)}$$

$$y_{k+1} = y_k \sqrt{\frac{2}{1 + \sqrt{1-[2^{-k}y_k]^2}}} \quad , \qquad \text{(II)}$$

$$y_{k+1} = y_k \sqrt{\frac{2y_k}{y_{k-1} + y_k}} \quad . \qquad \text{(III)}$$

It will be seen that formula (I), though mathematically correct, is not suitable for numerical evaluation. Severe cancellation results. See Essentials, Demonstration 1.3-4.

3. Flow Diagram

Since (III) produces the correct values much faster than (II), only the flow diagram for (III) is given.

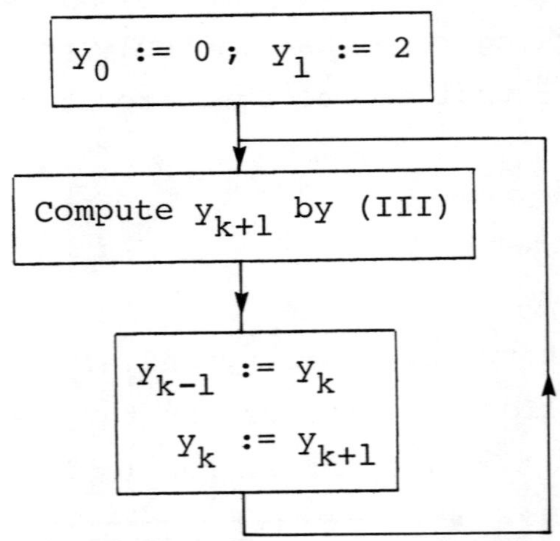

4. Storage and Program

(I)

R_0	R_1	R_2	R_3	R_4	R_5	R_6	R_7
y_k	2^k						

	00	10	20	30	40
0	⊠	x^2	STO 1		
1	2	−	*		
2	STO 0	$\sqrt{x}$	STO 0		
3	STO 1	−	R/S		
4	→ 1	2	GTO 04		
5	ENTER	÷			
6	1	$\sqrt{x}$			
7	RCL 0	RCL 1			
8	RCL 1	2			
9	÷	*			

(II)

R_0	R_1	R_2	R_3	R_4	R_5	R_6	R_7
y_k	2^k						

	00	10	20	30	40
0	⊠	x^2	STO 0		
1	2	−	R/S		
2	STO 0	$\sqrt{x}$	2		
3	STO 1	+	STO*1		
4	→ 1	2	GTO 04		
5	ENTER	$x \gtrless y$			
6	1	÷			
7	RCL 0	$\sqrt{x}$			
8	RCL 1	RCL 0			
9	÷	*			

(III) R_0 R_1 R_2 R_3 R_4 R_5 R_6 R_7

y_{k-1} y_k

	00	10	20	30	40
0	⊠	+			
1	CLX	÷			
2	STO 0	$\sqrt{x}$			
3	2	STO*1			
4	STO 1	RCL 1			
5	→ 2	R/S			
6	*	GTO 05			
7	RCL 0				
8	RCL 1				
9	STO 0				

5. Operating Instructions

These are the same for all three programs. Load program. Move operating switch to RUN and select mode of displaying numbers. Computation is started by pressing R/S; after a short time y_2 is displayed. After each display R/S must be pressed again in order to start new cycle. If only short displays of y_k are desired, the R/S instruction should be replaced by PAUSE.

6. Timing

for the computation of $y_2, \ldots, y_{11}$ (10 values), see Essentials, Demonstration 1.3-4, including short display

Program (I)	25 sec.
Program (II)	25 sec.
Program (III)	18 sec.

Demonstration 1.3-5:

RUNNING MEAN AND STANDARD DEVIATION OF A SEQUENCE OF DATA

1. Purpose

Given an open-ended sequence of real numbers $\{x_1, x_2, x_3, \ldots\}$, to compute, in a numerically stable manner, their running mean

$$\mu_n := \frac{1}{n} \sum_{k=1}^{n} x_k ,$$

and their standard deviation

$$\sigma_n := [\frac{1}{n} \sum_{k=1}^{n} (x_k - \mu_n)^2]^{1/2}$$

and to provide for the possibility of removing data points from the sequence.

2. Method

The most obvious method to compute μ_n and σ_n, based on accumulating the sums

$$s_n := \sum_{k=1}^{n} x_k , \qquad q_n := \sum_{k=1}^{n} x_k^2$$

is described by the following formulas

$$\mu_n = \frac{1}{n} s_n , \qquad \sigma_n^2 = \frac{1}{n} q_n - \mu_n^2 . \tag{I}$$

It can produce totally inaccurate values for σ_n^2 due to cancellation, See Essentials, Demonstration 1.3-5. Much more precise values of the standard deviation are obtained if the variances are computed recursively.

If s_n and σ_n^2 are known, and if a new value x_{n+1} is <u>added</u> to $\{x_1, x_2, \ldots, x_n\}$, the new values of mean and variance are given by

$$s_{n+1} = s_n + x_{n+1} , \qquad \mu_{n+1} = \frac{1}{n+1} s_{n+1} ,$$

$$\sigma_{n+1}^2 = \frac{n}{n+1} \sigma_n^2 + \frac{1}{n}(\mu_{n+1} - x_{n+1})^2 .$$

If the value x_n is <u>removed</u> from the data $\{x_1, x_2, \ldots, x_n\}$, the new values are

$$s_{n-1} = s_n - x_n , \qquad \mu_{n-1} = \frac{1}{n-1} s_{n-1} ,$$

$$\sigma_{n-1}^2 = \frac{n}{n-1} \sigma_n^2 - \frac{1}{n}(\mu_{n-1} - x_n)^2 .$$

Because the ordering of the data is clearly immaterial, the same formula applies if in place of x_n any of the values $x_1, \ldots, x_n$ is removed. Both sets of formulas given above may be combined into one set in the following way: Let $s := s_n$, $\mu := \mu_n$, $\sigma := \sigma_n$, let x denote the value being added to or removed from the data, and let s', μ', σ' denote the resulting new values of s, μ, σ. Also let

$$n' = \begin{cases} n+1 , & \text{if data is added} \\ n-1 , & \text{if data is removed} \end{cases}$$

Then

$$s' = s + (n' - n)x \ , \qquad \mu' = \frac{1}{n'} s' \ , \tag{IIa}$$

$$(\sigma')^2 = \frac{n}{n'} \sigma^2 + \frac{n' - n}{n} (\mu' - x)^2 \ . \tag{IIb}$$

On the HP-33E, s' and n' are computed automatically in either case by pressing $\sum+$ and $\sum-$ respectively.

3. Flow Diagram

Program I: See Essentials, Demonstration 1.3-5.

Program II: Upper signs refer to data being added, lower signs to removal of data.

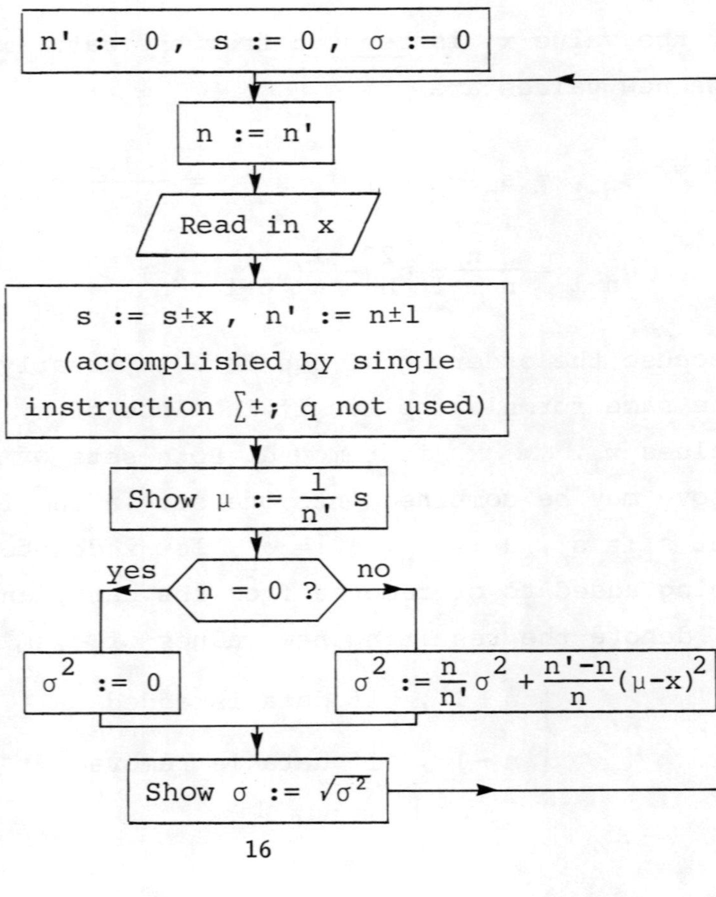

4. Storage and Program

(I)

R_0	R_1	R_2	R_3	R_4	R_5	R_6	R_7
		n	s	q	*	*	*

	00	10	20	30	40
0	⊠	RCL 4			
1	REG	RCL 2			
2	→ CLX	÷			
3	FIX 0	+			
4	R/S	$\sqrt{x}$			
5	FIX 8	PAUSE (R/S)			
6	$\bar{x}$	GTO 02			
7	PAUSE (R/S)				
8	x^2				
9	CHS				

(II)

R_0	R_1	R_2	R_3	R_4	R_5	R_6	R_7
σ^2	n	n'	s	*	*	*	*

	00	10	20	30	40
0	⊠	LAST X	*	GTO 02	
1	REG	-	RCL 0		
2	→ RCL 2	x^2	RCL 1		
3	STO 1	RCL 1	*		
4	CLX	x = 0	RCL 2		
5	FIX 0	GTO 27	÷		
6	R/S	÷	+		
7	FIX 8	RCL 2	→ STO 0		
8	$\bar{x}$	RCL 1	$\sqrt{x}$		
9	PAUSE (R/S)	-	PAUSE (R/S)		

5. Operating Instructions

The following instructions apply to both program (I) and program (II). Load program. Move operating switch to RUN. Press

PRGM
R/S

Calculator will display 0. This is a sign that data should be loaded into X-register. * After loading value of x, press

Σ+
R/S

if data is to be added or

Σ-
R/S

if data is to be removed. Calculator will briefly display new values of μ and σ, and stop by displaying 0. as an invitation to load new data. Return to *. - Display of μ and σ is in format chosen in step 05 of program (I) and 07 of program (II). This format may be chosen at will. If indeterminate displays of μ and σ are desired, change instructions 07 and 15 of program (I) or instructions 09 and 29 of program (II) to R/S. - The instructions Σ+ and Σ- simultaneously carry out several statistical operations (e.g. computation of q, see HP-33E-manual); this accounts for registers $R_4 \div R_7$ being used.

6. Examples

[1] See Essentials, Demonstration 1.3-5; program I, cancellation.

[2] See Essentials, Demonstration 1.3-5; given data $x_k = 1 + \varepsilon \frac{2k - m - 1}{m - 1}$, $k = 1, \ldots, m$, comparison of classical algorithm (I) with stable algorithm (II).

[3] Addition and removal of data.

	x_n	μ	σ (Program I)	σ (Program II)	σ (exact)
	100468	100468.0000	0.00000000	0.00000000	0.00000000
	100472	100470.0000	0.00000000	2.00000000	2.00000000
	100495	100478.3333	12.24744871	11.89773555	11.89771220
	100473	100477.0000	10.48808848	10.55937577	10.55935605
remove	100495	100471.0000	0.00000000	2.16037550	2.16024690
remove	100473	100470.0000	0.00000000	2.00020835	2.00000000
remove	100472	100468.0000	0.00000000	0.04082769	0.00000000

Problem 1.3-1:

SOLUTION OF CUBIC EQUATION BY SERIES EXPANSION

R_0	R_1	R_2	R_3	R_4	R_5	R_6	R_7
p	q	Σ	a_m	m	$-\frac{2q^2}{9p^3}$		

	00	10	20	30	40
0	╳	STO 2	STO+4	*	RCL 2
1	RCL 1	STO 3	RCL 4	1	RCL 3
2	2	STO*5	STO÷3	-	+
3	*	RCL 0	2	STO*3	STO 2
4	RCL 0	2	*	1	x = y
5	3	*	1	-	GTO 00
6	*	STO÷5	+	STO*3	GTO 19
7	÷	0	STO÷3	RCL 5	
8	STO 5	STO 4	RCL 4	STO*3	
9	CHS	→ 1	3	RCL 2	

Finds zero x of

$$x^3 + 3px + 2q = 0$$

from series expansion

$$x = -\frac{2q}{3p} \sum_{m=0}^{\infty} (-1)^m \frac{(3m)!}{m!(2m+1)!} \frac{(2q)^{2m}}{(3p)^{3m}}$$

(This is a special case of the Lagrange-Bürmann series, see ACCA I, chapter 1.)

Demonstration 1.4-1:

EVALUATION OF THE EXPONENTIAL SERIES

1. Purpose

To demonstrate smearing, see Essentials §1.4.

2. Method

We evaluate e^{-x}, $x > 0$, by summing the exponential series in floating point arithmetic, see Essentials, Demonstration 1.4-1 and §1.1. We stop summing as soon as two consecutive partial sums on the computer are equal. This criterion of termination is computer independent.

The following flow diagram and program apply to positive or negative values of x; hence the given program can also be used to evaluate e^{x}, $x > 0$, without smearing, see Essentials, Demonstration 1.4-1.

3. Flow Diagram

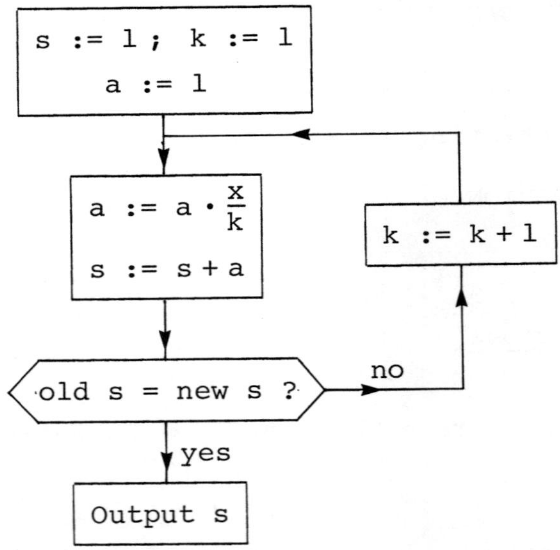

4. Storage and Program

R_0	R_1	R_2	R_3	R_4	R_5	R_6	R_7
x	s	a	k				
-x∈X							

	00	10	20	30	40
0		RCL 1	→ RCL 1		
1	STO 0	RCL 2	GTO 00		
2	1	RCL 1			
3	STO 1	+			
4	STO 2	STO 1			
5	STO 3	x = y			
6	→ RCL 0	GTO 20			
7	STO*2	1			
8	RCL 3	STO+3			
9	STO÷2	GTO 06			

5. Operating Instructions

Load program, move to RUN. Select mode of displaying numbers, e.g. by pressing

SCI 6

To compute e^{-x}, $x > 0$, load -x into X-register and press

PRGM
R/S

To obtain e^{-x} for other values of x load -x into X-register and press

R/S

By pressing

MANT

after computing e^{-x} all the digits that are computed internally can be made visible briefly (for longer display do not lift finger from MANT key).

6. Examples

[1] See Essentials, Demonstration 1.4-1

[2] Some values of e^{x}, $x > 0$; see also Essentials, Demonstration 1.4-1.

x	e^x	e^{-x} by $1/e^x$
1	2.718281830	$3.678794410*10^{-1}$
5	$1.484131593*10^{2}$	$6.737946990*10^{-3}$
10	$2.202646580*10^{4}$	$4.539992975*10^{-5}$
15	$3.269017376*10^{6}$	$3.059023202*10^{-7}$

Here only the last digit of each computed value is unreliable.

Problem 1.4-1:

BESSEL FUNCTION OF INTEGER ORDER

R_0	R_1	R_2	R_3	R_4	R_5	R_6	R_7
[n]	$\frac{x}{2}$	a_k	$\sum$	k	n!		

x∈X

	00	10	20	30	40
0		RCL 4	STO 3	STO÷2	GTO 24
1	2	x = 0	RCL 1	RCL 1	
2	÷	GTO 17	CHS	STO*2	
3	STO 1	STO*5	STO*1	RCL 3	
4	RCL 0	1	→ 1	RCL 3	
5	STO 4	STO-4	STO+4	RCL 2	
6	y^x	GTO 12	RCL 4	+	
7	STO 2	RCL 5	STO÷2	STO 3	
8	1	STO÷2	RCL 0	x = y	
9	STO 5	RCL 2	+	GTO 00	

Evaluates Bessel function $J_n(x)$ (n a nonnegative integer) from series

$$J_n(x) = \left(\frac{x}{2}\right)^n \sum_{k=0}^{\infty} (-1)^k \frac{\left(\frac{x}{2}\right)^{2k}}{k!(n+k)!}$$

Problem 1.4-2:

FUNCTIONS $f_n(x)$: EXPLICIT FORMULA

R_0	R_1	R_2	R_3	R_4	R_5	R_6	R_7
[n]	x	k	a_k	s			

x∈X

	00	10	20	30	40
0	☒	STO+4	STO÷4		
1	STO 1	1	RCL 3		
2	RCL 0	STO-2	STO-4		
3	STO 2	RCL 2	RCL 4		
4	STO 3	x = 0	CHS		
5	1	GTO 18	GTO 00		
6	STO 4	STO*3			
7	→ RCL 1	GTO 07			
8	STO*4	→ RCL 1			
9	RCL 3	e^x			

Evaluates

$$f_n(x) := \int_0^x t^n e^{-t} dt$$

by explicit formula

$$f_n(x) = n! - e^{-x}\{x^n + nx^{n-1} + n(n-1)x^{n-2} + \ldots + n!\} .$$

Problem 1.4-2:

FUNCTIONS $f_n(x)$: UNSTABLE SERIES

R_0	R_1	R_2	R_3	R_4	R_5	R_6	R_7
[n]	x	$\sum$	a_k	n+k+1	k		

$x \in X$

	00	10	20	30	40
0	⊠	STO÷3	STO÷3	y = x	
1	STO 1	STO 4	RCL 5	GTO 00	
2	RCL 0	0	STO÷3	STO 2	
3	1	STO 5	RCL 1	GTO 14	
4	+	→ RCL 4	CHS		
5	y^x	STO*3	STO*3		
6	STO 2	1	RCL 2		
7	STO 3	STO+5	RCL 2		
8	LAST X	+	RCL 3		
9	STO÷2	STO 4	+		

Evaluates

$$f_n(x) := \int_0^x t^n e^{-t} dt$$

by alternating series

$$f_n(x) := \sum_{k=0}^{\infty} (-1)^k \frac{x^{n+1+k}}{k!(n+k+1)}$$

Problem 1.4-2:

FUNCTIONS $f_n(x)$: STABLE SERIES

R_0	R_1	R_2	R_3	R_4	R_5	R_6	R_7
n (boxed)	x	a_k	$\sum$	n+k+1			

$x \in X$

	00	10	20	30	40
0	☒	RCL 1	STO*2		
1	STO 1	e^x	RCL 3		
2	RCL 0	STO÷2	RCL 3		
3	1	RCL 2	RCL 2		
4	STO 2	STO 3	+		
5	+	→ 1	STO 3		
6	STO 4	STO+4	x = y		
7	STO÷2	RCL 4	GTO 00		
8	y^x	STO÷2	GTO 15		
9	STO*2	RCL 1			

Evaluates

$$f_n(x) := \int_0^x t^n e^{-t} \, dt$$

by series of positive terms,

$$f_n(x) = \sum_{k=0}^{\infty} \frac{n! \, x^{n+k+1}}{(n+k+1)!}$$

Problem 1.4-3:

ERROR INTEGRAL, UNSTABLE SERIES

R_0	R_1	R_2	R_3	R_4	R_5	R_6	R_7
x	$\sum$	a_n	n	$-x^2$			

$x \in X$

	00	10	20	30	40
0		x^2	*	RCL 1	
1	STO 0	CHS	1	RCL 2	
2	2	STO 4	-	+	
3	*	0	STO*2	x = y	
4	π	STO 3	2	GTO 00	
5	$\sqrt{x}$	→ 1	+	STO 1	
6	÷	STO+3	STO÷2	GTO 15	
7	STO 1	RCL 3	RCL 4		
8	STO 2	STO÷2	STO*2		
9	RCL 0	2	RCL 1		

Evaluates

$$F(x) := \frac{2}{\sqrt{\pi}} \int_0^x e^{-t^2} dt$$

by series

$$F(x) = \frac{2}{\sqrt{\pi}} \sum_{n=0}^{\infty} (-1)^n \frac{x^{2n+1}}{(2n+1)n!}$$

Problem 1.4-3:

ERROR INTEGRAL, STABLE SERIES

R_0	R_1	R_2	R_3	R_4	R_5	R_6	R_7
x	$\sum$	a_n	$n+\frac{1}{2}$	x^2			

x∈X

	00	10	20	30	40
0	☒	π	RCL 3	GTO 00	
1	STO 0	$\sqrt{x}$	1	STO 1	
2	x^2	÷	+	GTO 18	
3	STO 4	STO 1	STO 3		
4	CHS	STO 2	STO÷2		
5	e^x	2	RCL 1		
6	RCL 0	1/x	RCL 1		
7	*	STO 3	RCL 2		
8	2	→ RCL 4	+		
9	*	STO*2	x = y		

Evaluates

$$F(x) := \frac{2}{\sqrt{\pi}} \int_0^x e^{-t^2} dt$$

by series

$$F(x) = \frac{2x}{\sqrt{\pi}} e^{-x^2} \sum_{n=0}^{\infty} \frac{x^{2n}}{(\frac{3}{2})_n}$$

Demonstrations 1.4-2, 1.5-1, 1.5-2, 1.5-3:

AN OBNOXIOUS FUNCTION TO EVALUATE

1. Purpose

To compute numerical values of the following functions y_n, defined by definite integrals,

$$y_n := \int_0^1 \frac{x^n}{x+a}\,dx$$

for a fixed value of $a > 1$ and for $n = 0, 1, 2, \ldots, 10$. The algorithm used should not be sensitive to smearing or numerical instability.

2. Method

The following explicit or recursion formulas present themselves for the computation of the values y_n

$$y_n = \sum_{k=0}^{n-1} (-1)^k a^k \binom{n}{k} [(1+a)^{n-k} - a^{n-k}] \frac{1}{n-k} + (-1)^n a^n \log \frac{1+a}{a}, \qquad \text{(I)}$$

$$y_0 = \log \frac{1+a}{a}; \quad y_n = \frac{1}{n} - ay_{n-1}, \quad n = 1,2,\ldots,10, \qquad \text{(II)}$$

$$y_{\text{"large"}n} := 0; \quad y_{n-1} = \frac{1}{a}\left(\frac{1}{n} - y_n\right), \qquad \text{(III)}$$

$$y_n = \sum_{k=0}^{\infty} \frac{(-1)^k}{(n+k+1)a^{k+1}} , \tag{IV}$$

see Essentials, Demonstrations 1.4-2, 1.5-1, 1.5-2 and 1.5-3. By programming each formula, it will be seen that only (III) and (IV) are reliable numerically, whereas (I) and (II) are "smeared" or unstable, respectively.

3. Flow Diagram

The flow diagrams corresponding to (I) and (II) will not be given explicitly. In (I) the sum (increasing k) is computed first, then the log-term is added; the binomial coefficient is computed recursively.

(III)

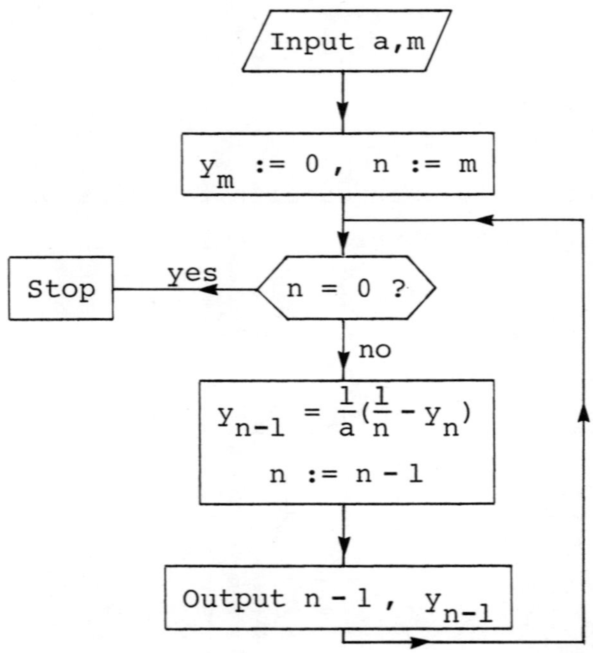

(IV)

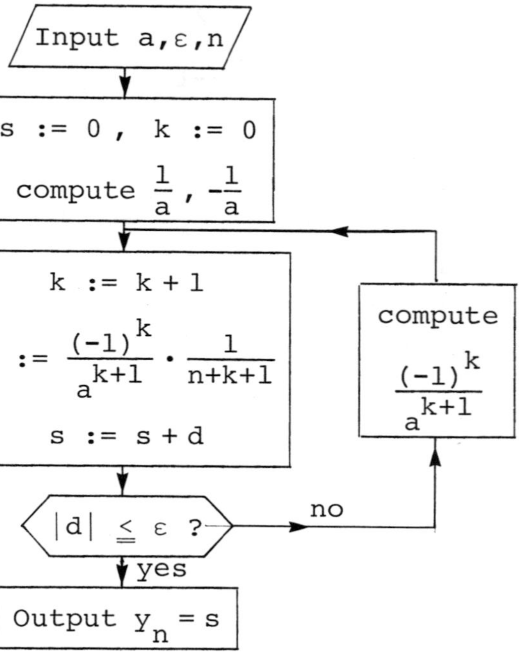

4. Storage and Program

(I)

R_0	R_1	R_2	R_3	R_4	R_5	R_6	R_7
[n]	[a]	n-k	k	$(-a)^k$	$\binom{n}{k}$	s	

	00	10	20	30	40
0	⊠	→ x = 0	-	STO*4	→ RCL 1
1	RCL 0	GTO 40	RCL 5	RCL 2	1
2	STO 2	RCL 1	*	STO*5	+
3	CLX	1	RCL 4	1	RCL 1
4	STO 6	+	*	STO-2	÷
5	STO 3	RCL 2	RCL 2	STO+3	ℓn
6	1	y^x	÷	RCL 3	RCL 4
7	STO 4	RCL 1	STO+6	STO÷5	*
8	STO 5	RCL 2	RCL 1	RCL 2	STO+6
9	RCL 2	y^x	CHS	GTO 10	RCL 6

(II) R_0 R_1 R_2 R_3 R_4 R_5 R_6 R_7

n [a] y_{n-1}

	00	10	20	30	40
0	⊠	STO 0	R/S		
1	RCL 1	→ RCL 1	GTO 11		
2	1	CHS			
3	+	STO*2			
4	RCL 1	RCL 0			
5	÷	1/x			
6	ℓn	STO+2			
7	STO 2	1			
8	R/S	STO+0			
9	1	RCL 2			

(III) R_0 R_1 R_2 R_3 R_4 R_5 R_6 R_7

[m] [a] y_n

n

	00	10	20	30	40
0	⊠	÷	GTO 03		
1	CLX	STO 2			
2	STO 2	1			
3	→ RCL 0	STO-0			
4	x = 0	RCL 0			
5	GTO 00	FIX 0			
6	1/x	PAUSE			
7	RCL 2	RCL 2			
8	-	FIX 9			
9	RCL 1	R/S			

(IV)

R_0	R_1	R_2	R_3	R_4	R_5	R_6	R_7
[n]	[a]	[ε]	$-\frac{1}{a}$	$\frac{(-1)^k}{a^{k+1}}$	s		
n+k+1							

	00	10	20	30	40
0	╳	RCL 2	GTO 08		
1	CLX	RCL 4	→ RCL 5		
2	STO 5	RCL 0	GTO 00		
3	RCL 1	÷			
4	1/x	STO+5			
5	STO 4	ABS			
6	CHS	$x \leq y$			
7	STO 3	GTO 21			
8	→ 1	RCL 3			
9	STO+0	STO*4			

5. Operating Instructions

Load one of the programs (I) ÷ (IV) and move to RUN; from now on they differ slightly for each program.

(I) Choose mode of displaying numbers, e.g. by pressing FIX 9. Then load n into R_0 and a into R_1. Upon pressing

PRGM
R/S

y_n will be displayed. To obtain another y_n, load new n into R_0 and press R/S etc.

(II) Here the y_n have to be determined in their natural order, hence only load a into R_1 and press

FIX 9

PRGM

R/S

y_0 will be displayed. To obtain y_k, $k \geqq 1$, press R/S after display of y_{k-1}.

(III) Here backward recursion is used from an input index m on downward. Hence load m into R_0 and a into R_1. Appropriate modes of displaying numbers are chosen automatically in the program. Press

PRGM

R/S

There results a brief display of $m - 1$ and an indefinite display of y_{m-1}. Press R/S to obtain $m - 2$ and y_{m-2}, etc.

(IV) The summation of the infinite series is stopped as soon as the next term to be added in absolute value is less than a given tolerance ε. Load n into R_0, a into R_1 and ε into R_2, e.g. if $\varepsilon = 10^{-12}$ by pressing

EEX

1

2

CHS

Then press

FIX 9

PRGM

R/S

to yield display of y_n. Load new n into R_0 and press R/S to obtain corresponding y_n, etc.

6. Examples

For (I) see Demonstration 1.4-2,
for (II) see Demonstration 1.5-1,
for (III) see Demonstration 1.5-2,
for (IV) see Demonstration 1.5-3.

Problem 1.5-1:

FUNCTIONS $f_n(x)$: FORWARD RECURRENCE

R_0	R_1	R_2	R_3	R_4	R_5	R_6	R_7
n	x	$e^{-x}x^n$	f_n				

$x \in X$

	00	10	20	30	40
0	☒	STO 3	RCL 3		
1	STO 1	R/S	R/S		
2	CHS	→ RCL 1	GTO 12		
3	e^x	STO*2			
4	STO 2	1			
5	0	STO+0			
6	STO 0	RCL 0			
7	1	STO*3			
8	RCL 2	RCL 2			
9	-	STO-3			

Evaluates

$$f_n(x) := \int_0^x t^n e^{-t} dt , \qquad n = 0, 1, 2, \ldots$$

by recurrence

$$f_0 = 1 - e^{-x} , \qquad f_n = nf_{n-1} - x^n e^{-x}$$

Problem 1.5-1:

FUNCTIONS $f_n(x)$: BACKWARD RECURRENCE

R_0	R_1	R_2	R_3	R_4	R_5	R_6	R_7
$\boxed{n_0}$	x	$x^n e^{-x}$	f_n				
$x \in X$							

	00	10	20	30	40
0	╳	→ RCL 2	RCL 3		
1	STO 1	STO+3	R/S		
2	RCL 0	RCL 0	RCL 1		
3	y^x	x = 0	STO÷2		
4	STO 2	GTO 00	GTO 10		
5	RCL 1	STO÷3			
6	e^x	1			
7	STO÷2	-			
8	0	STO 0			
9	STO 3	PAUSE			

Evaluates

$$f_n(x) := \int_0^x t^n e^{-t}\, dt$$

by recurrence

$$f^*_{n_0} = 0\ , \qquad f^*_{n-1} = \frac{1}{n}(f_n + x^n e^{-x})\ .$$

For n_0 sufficiently large, $f^*_n \approx f_n$.

Demonstration 1.7-5:

EVALUATION OF A NUMERICALLY ILL-BEHAVED FUNCTION

1. Purpose

To evaluate

$$f(x) := \begin{cases} 0 \quad , & x = 0 \\ \dfrac{1}{\sqrt[8]{\log \frac{1}{x}}} \quad , & x > 0 \end{cases}$$

2. Method

Straightforward.

3. Flow Diagram

Uninteresting, hence omitted.

4. Storage and Program

R_0	R_1	R_2	R_3	R_4	R_5	R_6	R_7
x∈X							

	00	10	20	30	40
0	╳	GTO 00			
1	FIX 8	→ 1/x			
2	x > 0	ℓn			
3	GTO 11	8			
4	x < 0	1/x			
5	GTO 08	CHS			
6	0	y^x			
7	GTO 00	GTO 00			
8	→ 0				
9	FIX 0				

5. Operating Instructions

Load program, move to RUN. Load value of x, $x \geqq 0$, into X-register. Upon pressing

PRGM
R/S

f(x) will be displayed. Load other value of x, if desired, and press

R/S

to obtain corresponding f(x), etc. If $x < 0$ is input

accidentally, the computer will display 0. (instead of error message which would spoil computation of next f(x)). The format is chosen in step 01.

6. Example

See Essentials, Demonstration 1.7-5.

Demonstration 1.7-15:

SOLUTION OF AN ILL-CONDITIONED INITIAL VALUE PROBLEM BY DISCRETIZATION

1. Purpose

To demonstrate the ill-conditioning of the initial value problem

$$y'' = y , \quad y(0) = 1 , \quad y'(0) = -1 ,$$

see Essentials, Demonstrations 1.7-8 and 1.7-15.

2. Method

Discretization, see Essentials, Demonstration 1.7-15.

3. Flow Diagram

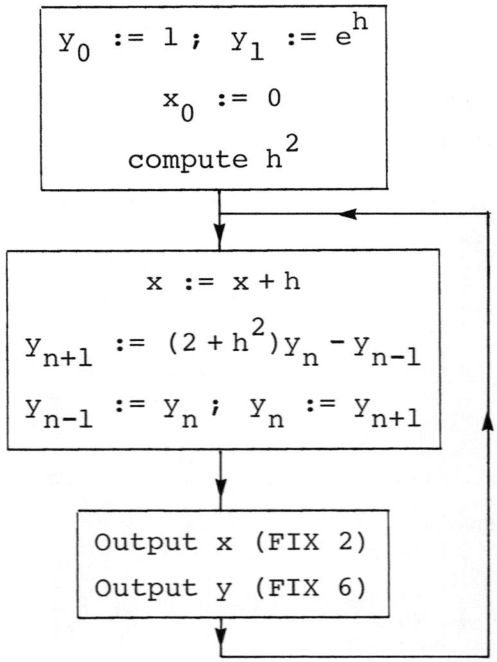

4. Storage and Program

R_0	R_1	R_2	R_3	R_4	R_5	R_6	R_7
[h]	y_{n-1}	y_n	h^2	x_n			

	00	10	20	30	40
0	☒	STO 4	*		
1	1	→ RCL 0	+		
2	STO 2	STO+4	STO 2		
3	RCL 0	RCL 1	RCL 4		
4	e^x	CHS	FIX 2		
5	STO 1	2	PAUSE		
6	RCL 0	RCL 3	RCL 2		
7	x^2	+	FIX 6		
8	STO 3	RCL 2	PAUSE		
9	0	STO 1	GTO 11		

5. Operating Instructions

Load program, move operating switch to RUN and load discretization step h into R_0. Upon pressing

PRGM

R/S

the computation is started. The computer will always first briefly display x_n (in easy-to-read FIX 2 format) and then briefly show the corresponding y_n (in FIX 6 format). If other modes of displaying x_n and y_n are desired, then instructions 24 and 27 have to be changed accordingly. By replacing instructions 25 and 28 by R/S the brief displays of x_n and y_n become stops; to restart the computation R/S needs to be pressed.

6. Example

See Essentials, Demonstration 1.7-15.

Demonstrations 2.1-1 ÷ 2.1-5:

ITERATION

1. Purpose

Computation of the fixed point of a function f.

2. Method

As limit of the iteration sequence generated by f, see Essentials §2.1, in particular Theorem 2.1. Two programs will be given:

(I) simply constructs $x_n = f(x_{n-1})$, $n = 1, 2, \ldots,$ without testing for convergence,

(II) stops as soon as two consecutive elements of the iteration sequence (on the computer) are equal, see Essentials, remarks at end of §2.1.

3. Flow Diagram

For (II) see Essentials, Fig. 2.1a with test "$n = n_{max}$?" removed.

4. <u>Storage and Program</u>

(I)	R_0	R_1	R_2	R_3	R_4	R_5	R_6	R_7
	n	x_{n-1}						

	00	10	20	30	40
0	⊠				
1	CLX				
2	STO 0				
3	→ 1				
4	STO+0				
5	GSB 07				
6	GTO 03				
7	→				
8					
9					

All the remaining programming spaces (and storage registers $R_2 \div R_7$) are available for the subroutine computing and displaying f(x); it should assume x in R_1, store f(x) into R_1 and end with the instruction RTN.

(II)	R_0	R_1	R_2	R_3	R_4	R_5	R_6	R_7
	n	$\boxed{x_{n-1}}$	x_n					

	00	10	20	30	40
0	☒	STO 1			
1	CLX	GTO 03			
2	STO 0	→ RCL 2			
3	1	GTO 00			
4	STO+0	→			
5	GSB 14				
6	RCL 1				
7	x = y				
8	GTO 12				
9	RCL 2				

All the remaining programming spaces (and storage registers $R_3 \div R_7$) are available for the subroutine computing and displaying f(x); unless each x_n needs to be registered, it is reasonable to use PAUSE for the display of f(x). The subroutine should assume x in R_1 and store f(x) into R_2 and end with the instruction RTN.

5. Operating Instructions

Load program (I) or (II) including subroutine to compute f(x). Move to RUN. Select mode of displaying numbers. Load starting value x_0 into R_1. Also load possible constants of the function f into appropriate registers. Press

PRGM

R/S

to start computation. In (I) the computer will go on indefinitely, provided that the f(x) within the subroutine are shown briefly (using PAUSE), otherwise R/S has to be pressed after each display; the convergence has to be checked by eye. In (II) the computer stops as soon as the criterion given in section 2 is satisfied; the number displayed is the fixed point of f. The number n of iteration steps can be obtained by pressing

RCL 0

6. Examples

[1] See Essentials, Demonstration 2.1-1: f(x) := cos x. Using (I) the soubroutine for f is:

```
07   RCL 1      09   cos        11   PAUSE
08   RAD        10   STO 1      12   RTN
```

[2] See Essentials, Demonstrations 2.1-2 and 2.1-5: $f(x) := \sqrt{2+x}$. Using (II) the subroutine for f is:

```
14   RCL 1      17   √x         20   RTN
15   2          18   STO 2
16   +          19   PAUSE
```

[3] See Essentials, Demonstrations 2.1-3 and 2.1-5: $f(x) := \frac{1}{a+x}$. Using (II) and assuming a in R_7 the subroutine for f is:

14	RCL 1	17	1/x	20	RTN
15	RCL 7	18	STO 2		
16	+	19	PAUSE		

With $x_0 = 0$ for a = 1 we get $s = x_{24} = 0.618033989$; if a = 1.000000001 the iteration sequence starts to cycle between 0.618033988 and 0.618033989.

Demonstrations 2.2-1, 2.2-3:

ITERATION WITH AITKEN ACCELERATION

1. Purpose

Computation of the fixed point of a function f more rapidly than with ordinary iteration.

2. Method

Aitken acceleration. Along with the sequence $\{x_n\}$ (see program "Iteration") the sequence $\{x_n'\}$ of accelerated values is generated:

$$x_n' := x_n + \frac{(x_{n+1} - x_n)^2}{(x_{n+1} - x_n) - (x_{n+2} - x_{n+1})} ,$$

see Essentials, §2.2.

3. Flow Diagram

See Essentials, §2.2.

4. Storage and Program

R_0	R_1	R_2	R_3	R_4	R_5	R_6	R_7
x_0	x_1	x_2					
$x_0 \in X$							

	00	10	20	30	40
0	╳	STO 1	RCL 1		
1	STO 0	GSB 28	-		
2	GSB 28	STO 2	-		
3	STO 1	→ RCL 1	÷		
4	GSB 28	RCL 0	RCL 0		
5	STO 2	-	+		
6	GTO 13	x^2	R/S		
7	→ RCL 1	LAST X	GTO 07		
8	STO 0	RCL 2	→		
9	RCL 2	PAUSE			

Program locations 28 ÷ 49 (and storage registers $R_3 \div R_7$) are available for the subroutine to compute f(x); it should assume x in X and also leave f(x) in the X-register (and end with the instruction RTN).

5. Operating Instructions

Load the program, including the subroutine to compute f. Switch to RUN and select the mode of displaying numbers, for instance by pressing

FIX 8

First press the starting value x_0 into the X-register and then

PRGM
R/S

to start the computation. The computer briefly displays x_{n+2} and stops at the display of x_n'. Press

R/S

to start new cycle. Convergence is not tested nor are iterations counted.

6. Examples

[1] See Essentials, Demonstration 2.2-1 and Program "Iteration", Example [1]: $f(x) := \cos x$. The subroutine for f is:

28	RAD	29	cos	30	RTN

[2] See Essentials, Demonstration 2.2-2 and Program "Iteration", Example [2]: $f(x) = \sqrt{2+x}$. The subroutine for f is:

28	2	30	$\sqrt{x}$
29	+	31	RTN

Demonstrations 2.2-3 ÷ 2.2-5:

AITKEN-STEFFENSEN ITERATION

1. Purpose

To compute the fixed points of smooth functions f without restrictions on the slope of f.

2. Method

Aitken-Steffensen iteration. The sequence $\{x_0^{(k)}\}$ is generated according to the following rule: Given an initial value $x_0 =: x_0^{(0)}$ * we construct $x_1 = f(x_0)$, $x_2 = f(x_1)$ and x_0' by Aitken's rule, see program "Iteration with Aitken acceleration". Now we use x_0' as starting point for a new iteration and go back to *. Convergence to the fixed point s takes place, provided that $f'(s) \neq 1$ and that x_0 is chosen "sufficiently close" to s, see Essentials §2.2.

3. Flow Diagram

See Essentials, §2.2.

4. Storage and Program

R_0	R_1	R_2	R_3	R_4	R_5	R_6	R_7
x_0	x_1	x_2					
	x_1-x_0	x_2-x_1					

$x_0 \in X$

	00	10	20	30	40
0	╳	RCL 1			
1	→ STO 0	x^2			
2	GSB 19	RCL 1			
3	STO 1	RCL 2			
4	GSB 19	−			
5	STO 2	÷			
6	RCL 1	+			
7	STO−2	R/S			
8	RCL 0	GTO 01			
9	STO−1	→			

Program locations 19 ÷ 49 (and storage registers $R_3 \div R_7$) are available for the subroutine to compute f(x); it should assume x in X and also leave f(x) in the X-register (and end with the instruction RTN).

5. Operating Instructions

They are the same as for the program "Iteration with Aitken acceleration". Each $x_0^{(k)}$ is displayed.

6. Examples

[1] See Essentials, Demonstration 2.2-3, Programs "Iteration" and "Iteration with Aitken Acceleration", Example [1]: $f(x) = \cos x$.

[2] See Essentials, Demonstration 2.2-4, Programs "Iteration" and "Iteration with Aitken Acceleration", Example [2]: $f(x) = \sqrt{2 + x}$.

[3] See Essentials, Demonstration 2.2-5: $f(x) = 10 \cos x$. The subroutine for f is:

```
19  RAD     21  1     23  *
20  cos     22  0     24  RTN
```

Section 2.2:

BISECTION

1. Purpose

To find a zero of a continuous function f, i.e. a solution of the equation

$$f(x) = 0$$

if points a and b are known such that

$$f(a) < 0 \quad \text{and} \quad f(b) > 0 .$$

(It is not required that $a < b$.)

2. Method

Successive bisection, see Essentials, §2.3. Let

$$x := \frac{a+b}{2} .$$

If $f(x) < 0$, then f has a zero between x and b; we thus set $a := x$ and repeat. If $f(x) \geqq 0$, then f becomes zero between a and x (or possibly at x); we thus set $b := x$ and repeat. The iteration is terminated if (on the machine) x = a or x = b. [It is not possible to use a = b as a convergence test; because of rounding errors this condition may never be satisfied.]

3. Flow Diagram

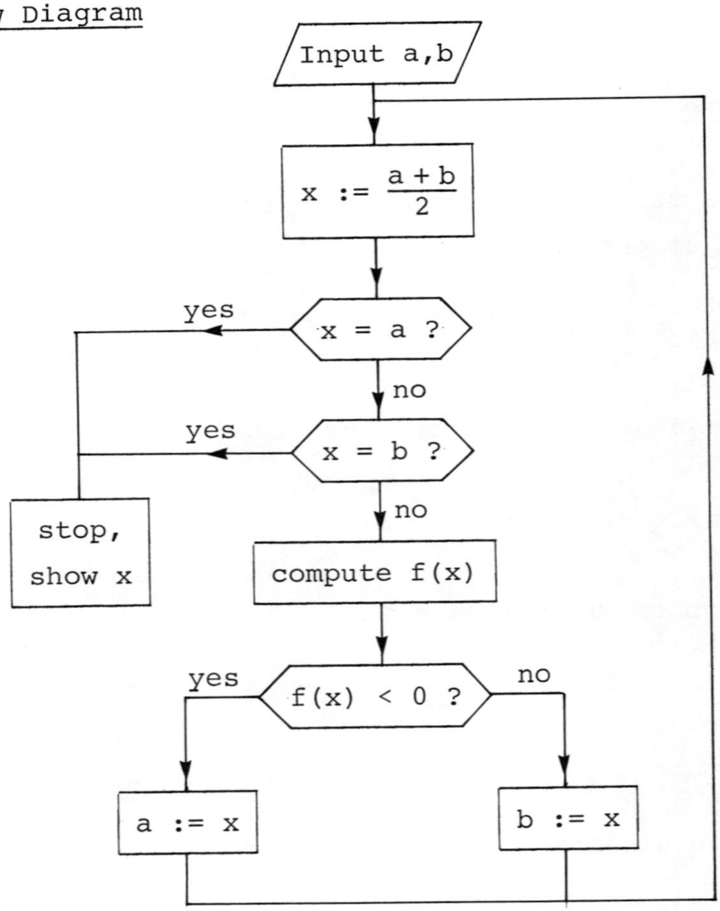

4. Storage and Program

R_0	R_1	R_2	R_3	R_4	R_5	R_6	R_7
a	b	x					

	00	10	20	30	40
0		RCL 1	GTO 01		
1	→ RCL 0	RCL 2	→ RCL 2		
2	RCL 1	x = y	STO 0		
3	+	R/S	GTO 01		
4	2	PAUSE	→		
5	÷	GSB 24			
6	STO 2	x < 0			
7	RCL 0	GTO 21			
8	x = y	RCL 2			
9	R/S	STO 1			

Locations 24 ÷ 49 are reserved for the subroutine to compute f(x). This program may assume x in the X-register, and also in R_2. The subroutine should put f(x) into the X-register (and its last instruction should be RTN). Registers $R_3 \div R_7$ are available for auxiliary storage.

5. Operating Instructions

Load program, including subroutine to compute f(x). (If f involves trigonometric functions, whose argument is measured in radians, include RAD instruction in subroutine.) Move operating switch to RUN position.

Select mode of displaying numbers. Load a with $f(a) < 0$ into R_0 and b with $f(b) > 0$ into R_1. (The domain of definition of f must contain the closed interval bounded by a and b.) Pressing

PRGM
R/S

starts computation. The computer briefly displays each iterate x and stops when (on the machine) $x = \frac{1}{2}(a+b)$ coincides with either a or b.

6. Examples and Timing

[1] Let $f(x) := \sin x$. We want to find the smallest positive zero of f by bisection. We know that it has the exact value $\pi = 3.1415926536...$ and therefore set $a := 4$ and $b := 1$. The subroutine for f is

24	RAD	25	sin	26	RTN

and we obtain $x = 3.141592654$. The error, as expected, is about 1 unit in the last digit displayed. Computing time 80 sec. Starting with $a := 4$ and $b := 3$ reduces the computing time by 5 sec.

[2] Let

$$f(x) := \operatorname{Log}\frac{1}{x} - x \operatorname{Arctan}\frac{1}{x} - \operatorname{Log} 2$$

(see ACCA III, §16.5). By experimentation we find $f(0.1) > 0$, $f(1) < 0$. We thus set $a := 1$, $b := 0.1$. Calculation with the following subroutine yields $x = 0.330587542$.

24	1/x	28	TAN^{-1}	32	2
15	LN	29	RCL 2	33	LN
26	LAST X	30	*	34	-
27	RAD	31	-	35	RTN

Computing time 135 sec.

[3] The seemingly trivial problem of finding the zero of $f(x) := x$ is not trivial for the bisection method as programmed here. If the absolute value of the solution is very small or zero, it takes many iterations to meet the criterion $x = a$ or $x = b$ due to the floating point representation used by the calculator. Thus in the present case, the iteration grinds down to values of $x < 10^{-99}$ before finally producing the correct solution $x = 0$. In compensation, the solution is very accurate. It can be guaranteed to be between $\pm 10^{-99}$. The same effect occurs, for instance, in the solution of $\sin x = 0$ for $a = -1$, $b = 2$.

Demonstrations 2.3-1, 2.3-2:

NEWTON'S METHOD: SQUARE ROOT ITERATION, DIVISION

1. Purpose

To apply Newton's method to the problems of finding the square root and the reciprocal of a given positive number.

2. Method

Newton's method, see Essentials, §2.3. The Newton iteration function to determine is

a) $\sqrt{c}$, $c > 0$ is $f(x) = \frac{1}{2}(x + \frac{c}{x})$

b) $\frac{1}{a}$, $c > 0$ is $f(x) = x(2-ax)$,

see Demonstration 2.3-1 and 2.3-2. The iteration is stopped as soon as two consecutive iterates differ by less than ε, the tolerance ε been chosen such that full machine accuracy is obtained, hence for the HP-33E $\varepsilon := 10^{-10}$.

3. Flow Diagram

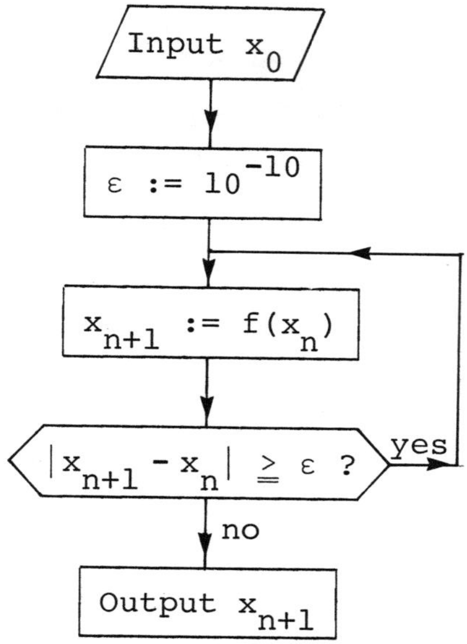

4. Storage and Program

R_0	R_1	R_2	R_3	R_4	R_5	R_6	R_7
c or a	x_0						
	x_n	ε					

	00	10	20	30	40
0	☒	STO 1			
1	1	-			
2	EEX	ABS			
3	1	RCL 2			
4	0	$x \leqq y$			
5	CHS	GTO 07			
6	STO 2	RCL 1			
7	→ RCL 1	R/S			
8	PAUSE	→			
9	GSB 18				

Locations 18 and following are reserved for the subroutine computing values of the iteration function. It should assume x in R_1 and c or a in R_0 and then leave f(x) in the X-register.

5. Operating Instructions

Load program including appropriate subroutine for the iteration function. Move to RUN. Load constant c or a into R_0 and starting value x_0 into R_1. Upon pressing

FIX 8
PRGM
R/S

the computation is started. The computer will briefly display each iterate (including x_0) until convergence has taken place, when it stops by displaying the result, i.e. the square root or the reciprocal. The iterations are not counted.

6. Examples

[1] Square root, see Demonstration 2.3-1. The subroutine for the Newton iteration function is:

18	RCL 0	21	RCL 1	24	÷
19	RCL 1	22	+	25	RTN
20	÷	23	2		

[2] Reciprocal, see Demonstration 2.3-2.
Subroutine:

```
18  RCL 0
19  RCL 1
20  *
21  CHS
22  2
23  +
24  RCL 1
25  *
26  RTN
```

Problem 2.3-4:

DRIVING MECHANISM

R_0	R_1	R_2	R_3	R_4	R_5	R_6	R_7
L	R	r	x	π(R+r)	R-r		

	00	10	20	30	40
0	⊠	÷	÷	*	RCL 3
1	RCL 0	STO 3	$\sin^{-1}$	RCL 3	x = y
2	RCL 1	RCL 1	RCL 5	x^2	GTO 00
3	RCL 2	RCL 2	*	RCL 5	x $\gtrless$ y
4	+	-	2	x^2	STO 3
5	π	STO 5	*	-	PAUSE
6	*	RAD	RCL 4	$\sqrt{x}$	GTO 17
7	STO 4	→ RCL 0	+	2	
8	-	RCL 5	-	*	
9	2	RCL 3	RCL 3	÷	

Solves the equation g(x) = L, where

$$g(x) = \pi(R+r) + 2\sqrt{x^2-(R-r)^2} + 2(R-r) \arcsin \frac{R-r}{x}$$

by Newton's method. Iteration function

$$f(x) := x - \frac{g(x) - L}{g'(x)}$$

is evaluated as

$$f(x) = x \frac{L - \pi(R+r) - 2(R-r) \arcsin \frac{R-r}{x}}{2\sqrt{x^2 - (R-r)^2}} .$$

Starting value is $x_0 = \frac{1}{2}(L - \pi(R+r))$.

Demonstrations 2.3-3, 2.3-4:

HORNER ALGORITHM

1. Purpose

Given an arbitrary polynomial of degree $\leqq 4$ with real coefficients

$$p(x) = a_0x^4 + a_1x^3 + a_2x^2 + a_3x + a_4 ,$$

and given an arbitrary real number x_0, to determine the coefficients b_m in the representation of p in powers of $h := x - x_0$,

$$p(x_0 + h) = b_0h^4 + b_1h^3 + b_2h^2 + b_3h + b_4 ,$$

that is, the Taylor coefficients of p at x_0.

2. Method

The Horner algorithm, see Essentials, §2.3.

3. Flow Diagram

Because an address modification is not available, the programming here is different from what it would be on a larger computer. By a cyclic permutation (here called rotation) of the a_k the coefficients operated on are always found in the same registers. After the algorithm is executed the Taylor coefficients b_k are stored where previously the coefficients a_k of the given polynomial were stored. A triangular array of coefficients has to be generated row by row; n and m are the row and column indices respectively.

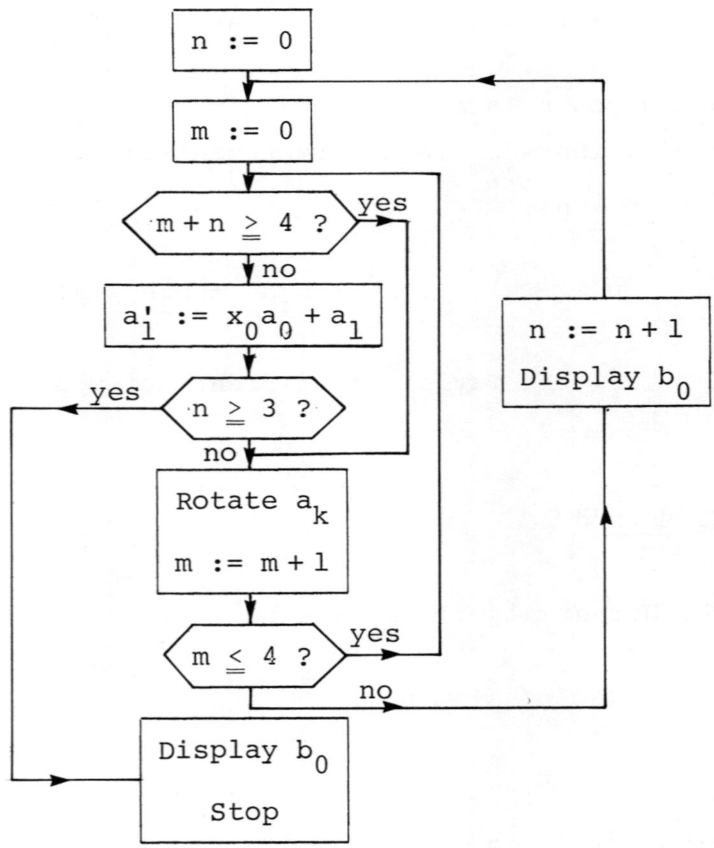

4. Storage and Program

R_0	R_1	R_2	R_3	R_4	R_5	R_6	R_7
a_0	a_1	a_2	a_3	a_4	x_0	*	n.m

	00	10	20	30	40
0		4	3	STO 2	RCL 7
1	→ RCL 7	$x \leqq y$	$x \leqq y$	RCL 4	FRAC
2	INT	GTO 23	GTO 49	STO 3	$x \leqq y$
3	STO 7	RCL 0	→ RCL 0	RCL 6	GTO 04
4	→ RCL 7	RCL 5	STO 6	STO 4	RCL 0
5	FRAC	*	RCL 1	•	PAUSE (R/S)
6	9	RCL 1	STO 0	1	1
7	*	+	RCL 2	STO+7	STO+7
8	RCL 7	STO 1	STO 1	•	GTO 01
9	+	RCL 7	RCL 3	4	→ RCL 0

Note: A fractional index n.m is used to save storage.

5. Operating Instructions

Load the program and switch to RUN. Load the coefficients a_m into R_m ($m = 0, 1, \ldots, 4$). Also load x_0 into R_5 and initial value zero of fractional index n.m into R_7. When

PRGM
R/S

is pressed, the calculator computes the b_m; it briefly displays b_0 after each Horner step and stops by displaying

b_0 after all the b_m have been computed. At this time, the b_m are stored in R_m (m = 0, 1, ..., 4,), thus enabling the operator to continue the computation immediately with a different x_0 (and after re-setting the index n.m equal to zero!). The original a_n are lost. If all the computed coefficients need to be recorded or if less than four Horner steps are desired, then instruction 45 should be changed to R/S, effecting a stop after each Horner step. At this time the coefficients can be found in corresponding registers as indicated above. Press R/S to continue the computation.

6. Examples and Timing

[1] $p(x) := x^4 - 4x^3 + 3x^2 - 2x + 5$. To compute the Taylor coefficients at $x = 2$, see Essentials, Demonstration 2.3-3. Computing time including short displays 38 sec.

[2] $p(x) := x^4 - 2x^3 + 8x - 16$, $p(-2) = 0$. To divide p by $x - (-2)$, see Essentials, Demonstration 2.3-4.

Demonstration 2.3-5:

NEWTON'S METHOD FOR POLYNOMIALS WITH AUTOMATIC DEFLATION

1. Purpose

To determine all real zeros of a real polynomial of degree 4,

$$p(x) := a_0x^4 + a_1x^3 + a_2x^2 + a_3x + a_4 .$$

2. Method

Newton's method for polynomials, see Essentials, §2.3.

3. Flow Diagram

This is identical with the flow diagram Figure 2.3. of Essentials, §2.3 for $n = 4$. For reasons of space the program given in section 4 does not check convergence and shows the new zero found only after the deflation has been performed.

4. Storage and Program

R_0	R_1	R_2	R_3	R_4	R_5	R_6	R_7
a_0	a_1	a_2	a_3	a_4	b	c	x

	00	10	20	30	40
0		RCL 3	→ STO+5	*	RCL 3
1	→ 0	GSB 20	RCL 7	STO+1	STO 4
2	STO 5	RCL 4	STO*6	RCL 1	RCL 2
3	STO 6	STO+5	RCL 5	RCL 7	STO 3
4	RCL 0	RCL 6	STO+6	*	RCL 1
5	GSB 20	STO÷5	RCL 7	STO+2	STO 2
6	RCL 1	RCL 5	STO*5	RCL 2	RCL 0
7	GSB 20	PAUSE	RTN	RCL 7	STO-0
8	RCL 2	STO-7	RCL 0	*	STO 1
9	GSB 20	GTO 01	RCL 7	STO+3	RCL 7

5. Operating Instructions

Load the program and then turn the operating switch to RUN. Select the mode of displaying numbers, for instance by pressing

FIX 8

Load the coefficients a_i of the given polynomial into R_i. [a_0 is the coefficient of x^4 etc.!] Load the starting value x_0 into R_7 (if $x_0 = 0$, it is already there). Press

PRGM

* R/S

to start computation. The computer will briefly display the corrections $-\Delta x_n = \frac{p(x_n)}{p'(x_n)}$ [For indeterminate display change instruction 17 to R/S and press R/S after each display.] Their convergence has to be checked by eye. As soon as they are zero within the accuracy desired

R/S

should be pressed. At this point the zero can be found in R_7. But after pressing

GTO 28
R/S

the polynomial will be deflated automatically, and the computer will halt while displaying the zero just found. In the process of deflation the new coefficients are written over the old ones. To compute the next zero load new starting value into R_7 (if this is omitted the last zero will be used as starting value) and repeat the process beginning with instruction * above. This may be done until all real zeros have been found.

6. Examples and Timing

1 Let $p(x) := p_4(x) := x^4 - 16x^3 + 72x^2 - 96x + 24$, the Laguerre polynomial of degree four. To determine all its zeros using Newton's method, see Essentials, Demonstration 2.3-5.

[2] $p(x) := (x-1)(x+2)(x-3)(x-4)$

$$= x^4 - 6x^3 + 3x^2 + 26x - 24$$

Starting with $x_0 = 0$ we get the zero of smallest absolute value

$$x_1 = 1.00000000$$

after five iterations. Starting again with $x_0 = 0$ - deflation is done automatically by the program - after ten iterations the computer produces the zero

$$x_2 = 4.00000000$$

[Upon looking at the graph of p it can be understood, that not the zeros -2 and 3 with smaller absolute values have been determined first.] Again using $x_0 = 0$ as starting value we get

$$x_3 = -2.00000000$$

after six iterations. After one more iteration we find - independently of the starting value now -

$$x_4 = 3.00000000$$

Computing time about 4 sec. per iteration.

3 Let $p(x) := E_4(x) = x^4 - 2x^3 + x$, the fourth Euler polynomial. Obviously

$$x_1 = 0$$

is a zero of p. Hence the zeros of

$$q(x) := x^3 - 2x^2 + 1$$

remain to be determined. Since $x = 0$ is a zero of $q'(x)$ we have to choose a nonzero starting value, e.g. $x_0 = .5$. After one iteration we obtain

$$x_2 = 1.00000000$$

Starting with $x_0 = 0$ after five iterations produces the zero

$$x_3 = -0.61803399$$

and in one more iteration

$$x_4 = 1.61803399$$

Computing time about 4 sec. per iteration.

4 Let $p(x) := T_8(x) = 128x^8 - 256x^6 + 160x^4 - 32x^2 + 1$, the Chebyshev polynomial of degree 8. To determine its zeros we first put $y := x^2$ and look for the zeros y_i, $i = 1, 2, 3, 4$, of

$$q(y) := 128y^4 - 256y^3 + 160y^2 - 32y + 1$$

The zeros of T_8 will then be the roots of y_i, $i = 1, 2, 3, 4$. In this example we will always use $x_0 = 0$ as starting value. We obtain

$y_1 = 0.03806023$ after 4 iterations

$y_2 = 0.30865828$ after 6 iterations

$y_3 = 0.69134172$ after 6 iterations

$y_4 = 0.96193977$ after 1 iteration

Hence the zeros of T_8 are

$$x_{1,2} = \pm 0.19509032$$

$$x_{3,4} = \pm 0.55557023$$

$$x_{5,6} = \pm 0.83146961$$

$$x_{7,8} = \pm 0.98078528$$

and lie between -1 and +1 as expected. Computing time about 5 sec. per iteration.

Demonstration 2.4-1:

ITERATION FOR SYSTEMS OF TWO EQUATIONS

1. Purpose

To compute the solution of

$$\begin{cases} x = f(x,y) \\ y = g(x,y) \end{cases}$$

2. Method

Iteration, see Essentials, §2.4.

3. Flow Diagram

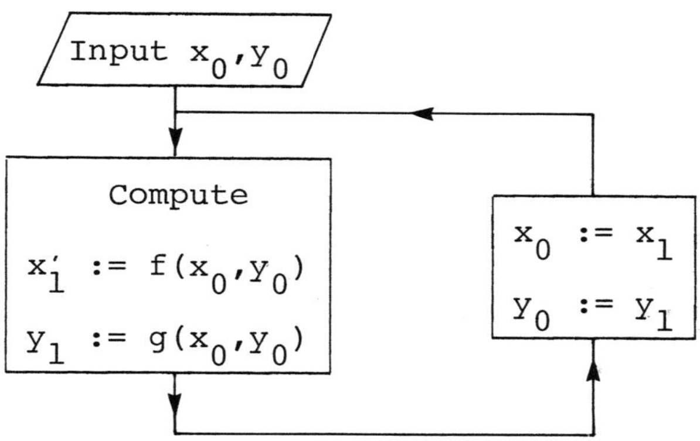

4. Storage and Program

R_0	R_1	R_2	R_3	R_4	R_5	R_6	R_7
x_0	y_0	*				A	B

	00	10	20	30	40
0	☒	→ RAD	RTN	+	
1	→ GSB 10	RCL 0	→ RAD	RTN	
2	STO 2	cos	RCL 0		
3	R/S	RCL 6	sin		
4	GSB 21	*	RCL 6		
5	STO 1	RCL 1	*		
6	R/S	sin	RCL 1		
7	RCL 2	RCL 7	cos		
8	STO 0	*	RCL 7		
9	GTO 01	-	*		

The foregoing program implements the example given below. If it is to be applied to another system, then use instructions 10 ÷ 49 for the two subroutines computing f and g and change instructions 01 and 04 accordingly. Storage locations $R_3 \div R_7$ are available for the computation of f and g.

5. Operating Instructions

Load program, including subroutines to compute f and g (see remark in section 4). Then switch to RUN and choose mode of displaying numbers. Load starting values x_0 and

y_0 into R_0 and R_1 respectively. Load constants for the computation of the functions f and g. Upon pressing

PRGM
R/S

the computation is started. The computer will stop by displaying x_1. Press R/S to obtain y_1 and again R/S to obtain x_2, etc. Convergence has to be tested by eye.

6. Example

Let $f(x,y) := A\cos x - B\sin y$

$g(x,y) := A\sin x - B\cos y$,

A,B constants, see Essentials, Demonstration 2.4-1.

Problem 2.4-1:

CIRCLE TOUCHING THREE GIVEN CIRCLES

R_0	R_1	R_2	R_3	R_4	R_5	R_6	R_7
x	y	a	b	r_1-r_2	r_1-r_3	x^2	d_1

	00	10	20	30	40
0	⊠	RCL 0	*	RCL 1	RCL 3
1	→ RCL 0	RCL 2	RCL 2	RCL 3	÷
2	x^2	-	÷	-	RCL 3
3	STO 6	x^2	RCL 2	x^2	+
4	RCL 1	RCL 1	+	RCL 6	2
5	x^2	x^2	2	+	÷
6	+	+	÷	$\sqrt{x}$	STO 1
7	$\sqrt{x}$	$\sqrt{x}$	STO 0	RCL 7	PAUSE
8	STO 7	RCL 7	PAUSE	+	GTO 01
9	RCL 4	+	RCL 5	*	

Let the centers and radii of the three given circles be (0,0), r_1; (a,0), r_2; (0,b), r_3. The center (x,y) of a circle touching the three circles, but not containing any of them, will satisfy

$$d_1 + r_1 = d_2 + r_2 ,$$

$$d_1 + r_1 = d_3 + r_3 ,$$

where $d_1 := \sqrt{x^2 + y^2}$, $d_2 := \sqrt{(x-a)^2 + y^2}$, $d_3 := \sqrt{x^2 + (y-b)^2}$.
These equations may be cast in the form

$$x = \frac{1}{2}\left\{ a + \frac{1}{a}(r_1 - r_2)(d_1 + d_2) \right\}$$

$$y = \frac{1}{2}\left\{ b + \frac{1}{b}(r_1 - r_3)(d_1 + d_3) \right\}$$

Program tries to solve these equations by iteration.
Iteration will converge if

$$\frac{|r_1 - r_2|}{a}, \qquad \frac{|r_1 - r_3|}{b}$$

are sufficiently small.

Problem 2.4-3:

SOLUTION OF 2 × 2 SYSTEM WITH SUCCESSIVE OVERRELAXATION

R_0	R_1	R_2	R_3	R_4	R_5	R_6	R_7
a	b	c	d	x_0	y_0	ω	
				x_n	y_n		

	00	10	20	30	40
0		-	-	PAUSE	-
1	RCL 2	RCL 4	RCL 6	RCL 2	RCL 1
2	RCL 0	*	*	RCL 4	RCL 4
3	RCL 5	+	1	-	*
4	*	STO 4	RCL 6	RCL 0	-
5	-	PAUSE	-	RCL 5	→ P
6	RCL 6	RCL 3	RCL 5	*	PAUSE
7	*	RCL 1	*	-	PAUSE
8	1	RCL 4	+	RCL 3	GTO 01
9	RCL 6	*	STO 5	RCL 5	

Solves system

$$x + ay = c$$

$$bx + y = d$$

by iterating in the form

$$x_{n+1} = \omega(c - ay_n) + (1 - \omega)x_n$$

$$y_{n+1} = \omega(d - bx_{n+1}) + (1 - \omega)y_n$$

(ω = overrelaxation parameter). Also shows length of residual vector $\underline{r}$,

$$|\underline{r}| = \sqrt{(c - x_n - ay_n)^2 + (d - bx_n - y_n)^2}$$

after each step.

Problem 2.5-1:

DELAYED EXPONENTIAL POPULATION GROWTH

R_0	R_1	R_2	R_3	R_4	R_5	R_6	R_7
$\boxed{\alpha}$	x	y	r	ϕ			

k∈X

	00	10	20	30	40
0		STO 2	RCL 0	→ P	RCL 1
1	RAD	RCL 0	+	RCL 0	-
2	4	$x \gtrless y$	→ P	*	STO+1
3	*	÷	STO 3	RCL 3	PAUSE
4	1	ℓn	R ↓	÷	$x \gtrless y$
5	-	STO 1	STO 4	$x \gtrless y$	RCL 2
6	2	RCL 2	RCL 2	RCL 4	-
7	÷	RCL 1	RCL 1	-	STO+2
8	π	e^x	1	$x \gtrless y$	PAUSE
9	*	→ R	+	→ R	GTO 16

Finds complex solution z of

$$z = \alpha e^{-z} \qquad (*)$$

by applying Newton's method to system of equations obtained by separating real and imaginary parts, starting with asymptotic solution value

$$z_0 = z_0^{(k)} := \mathrm{Log}\,\frac{2\alpha}{(4k-1)\pi} + i\,\frac{4k-1}{2}\pi ,$$

where $k = 1, 2, \ldots$. Corrections $\Delta x, \Delta y$ are shown; after convergence to sufficient accuracy has been achieved, solution is $z = x + iy$ where $x \in R_1$, $y \in R_2$. Starting value

$$z_0 = z_0^{(k)} = \mathrm{Log}\,\frac{2\alpha}{(4k-1)\pi} + i\,\frac{4k-1}{2}\pi ,$$

where $k = 1, 2, \ldots$ yields k-th complex root.

Algebraic simplification in Newton's equations results in iteration

$$z_{n+1} = f(z_n)$$

where

$$f(z) := \frac{z+1}{e^z + \alpha}$$

Use is made of polar representation of complex numbers to evaluate f.

Problems 2.5-2, 2.5-3:

EVALUATION OF REAL POLYNOMIAL OF DEGREE $\leqq$ 5
AT A COMPLEX POINT

R_0	R_1	R_2	R_3	R_4	R_5	R_6	R_7
a_0	a_1	a_2	a_3	a_4	a_5	r	θ

	00	10	20	30	40
0		RCL 6	RCL 6	RCL 6	RCL 6
1	RAD	*	*	*	*
2	RCL 7	$x \gtrless y$	$x \gtrless y$	$x \gtrless y$	$x \gtrless y$
3	RCL 6	RCL 7	RCL 7	RCL 7	RCL 7
4	RCL 0	+	+	+	+
5	*	$x \gtrless y$	$x \gtrless y$	$x \gtrless y$	$x \gtrless y$
6	→ R	→ R	→ R	→ R	→ R
7	RCL 1	RCL 2	RCL 3	RCL 4	RCL 5
8	+	+	+	+	+
9	→ P	→ P	→ P	→ P	→ P

Evaluates

$$p(z) = a_0 z^5 + a_1 z^4 + a_2 z^3 + a_3 z^2 + a_4 z + a_5$$

$$= ((((a_0 z + a_1)z + a_2)z + a_3)z + a_4)z + a_5$$

where the a_i are real and z is complex, $z = re^{i\theta}$. Data are preserved. Result appears in polar form, $p(z) = se^{i\alpha}$, where $s \in X$, $\alpha \in Y$. If display in cartesian form $u + iv$ is desired, change instruction 49 to GTO 00.

Demonstrations 3.2-1 ÷ 3.2-3:

MODIFIED BERNOULLI METHOD FOR REAL OR FOR COMPLEX CONJUGATE ZEROS

1. Purpose

To determine the one real zero or the two complex conjugate zeros of smallest modulus of a real polynomial of degree 4,

$$p(z) := a_0 z^4 + a_1 z^3 + a_2 z^2 + a_3 z + a_4 ,$$

provided that the remaining zeros have larger absolute values.

2. Method

The modified Bernoulli method using balanced starting values, see Essentials, §3.2. We will also apply this method to polynomials with multiple zeros and see that the convergence is not impaired. The key definitions and formulas for $k = 4$ used in the program given below are the following

$$\hat{q}_n := \frac{x_n}{x_{n+1}}$$

Balanced starting values:

$$\hat{q}_j = -\frac{a_4}{a_3 + \hat{q}_{j-1}(a_2 + \hat{q}_{j-2}(a_1 - (j+1)\frac{a_4}{a_3}a_{4-j-1}))}$$

$j = 1, 2, 3,$ where $\hat{q}_{-1} := \hat{q}_0 := 0$

Real zero z_1:

$$\hat{q}_k = -\frac{a_4}{a_3 + \hat{q}_{k-1}(a_2 + \hat{q}_{k-2}(a_1 + q_1 a_0))},$$

then shift $\hat{q}_j := \hat{q}_{j+1}$, $j = 1, 2, 3.$

$z_1 = \lim \hat{q}_n$; deflate with $z - z_1$ and apply the method to the deflated polynomial.

Complex conjugate zeros z_1, z_2:

$\hat{q}_k$ as in case of real zero

$$\hat{q}'_{k-1} := \frac{\hat{q}_{k-1} - \hat{q}_{k-2}}{\hat{q}_k - \hat{q}_{k-1}}\hat{q}_{k-1}$$

$$\hat{b}_n := \hat{q}_{k-1} + \hat{q}'_{k-1}, \qquad \hat{c}_n := \hat{q}_k \hat{q}'_{k-1}$$

$$\hat{b} := \lim \hat{b}_n, \qquad \hat{c} := \lim \hat{c}_n$$

z_1 and z_2 are the solutions of $z^2 - \hat{b}z + \hat{c} = 0$; deflate with this quadratic factor and apply the method to the deflated polynomial.

3. Flow Diagram

See Essentials, §3.2, Figure 3.2.

4. Storage and Program

R_0	R_1	R_2	R_3	R_4	R_5	R_6	R_7
a_0	a_1	a_2	a_3	a_4	$-2\frac{a_4}{a_3}$	0	0
					$\hat{q}_{k-1}$	$\hat{q}_{k-2}$	$\hat{q}_{k-3}$

	00	10	20	30	40
0		STO*7	STO 7	CHS	÷
1	GSB 13	→ GSB 13	*	STO 5	RCL 6
2	2	GTO 11	RCL 2	GTO ..	*
3	STO÷6	→ RCL 4	+	RCL 6	+
4	3	RCL 0	RCL 5	RCL 6	PAUSE
5	STO*6	RCL 7	STO 6	RCL 7	LAST X
6	GSB 13	*	*	-	RCL 5
7	3	RCL 1	RCL 3	RCL 5	*
8	STO÷7	+	+	RCL 6	PAUSE
9	4	RCL 6	÷	-	RTN

Real root: 32 GTO 48

Complex root: 32 GTO 33

5. Operating Instructions

Load program, where instruction 32 discriminates between the real and the complex case (see section 4). Switch to RUN. Load the coefficients a_i of the given polynomial into R_i, $i = 0, 1, 2, 3, 4$ (a_i is the coefficient of z^{4-i}). * Also load $-2\frac{a_4}{a_3}$ into R_5 and zero into R_6 and R_7. Press

PRGM

R/S

to start the computation. The computer will briefly display $\hat{q}_n$ (real root) or first $\hat{b}_n$ and then $\hat{c}_n$ (complex roots). The convergence of the $\hat{q}_n$ or $\hat{b}_n$ and $\hat{c}_n$ has to be checked by eye; if an indeterminate display of these quantities is desired change instruction 48 or instructions 44 and 48 to R/S and press R/S after each display to continue the computation. If the program should be used to compute further zeros of the polynomial, then it should be deflated with the corresponding linear or quadratic factor. This deflation has to be done by hand. If the coefficients of the deflated polynomial again are called $a_0 \div a_4$, then $a_0 = 0$ in the real case and $a_0 = a_1 = 0$ in the complex case. Before restarting the computation go to * above, etc.

6. Examples

[1] $p(z) = z^4 - 16z^3 + 72z^2 - 96z + 24$, see Essentials, Demonstration 3.2-1. We apply the program for real roots. In order to deflate with $z - z_1$ we perform one Horner step by hand:

```
RCL 0
RCL 5
*
STO+1
RCL 1
RCL 5
*
STO+2
RCL 2
RCL 5
*
STO+3
RCL 3
RCL 5
*
STO+4
```

At this point the coefficients of the deflated polynomial of degree 3 are located in $R_0 \div R_3$. The content of R_4 is equal to $p(z_1)$ and therefore should equal zero (up to rounding errors). We now have to put these coefficients into the appropriate storage registers and load $R_5 \div R_7$ accordingly:

```
RCL 3
STO 4
RCL 2
STO 3
RCL 1
STO 2
RCL 0
STO 1
0
STO 0
STO 6
STO 7
RCL 4
RCL 3
÷
2
*
CHS
STO 5
```

We now are ready to do the next Bernoulli real step. After the next zero has been found in this manner, we deflate and re-store as above etc.

2 $p(z) = (z-1)^3(z+2) = z^4 - z^3 - 3z^2 + 5z - 2$, see Essentials, Demonstration 3.2-2. Since $(z-1)^3 = (z-1)(z-)^2$, it makes sense to try to find one zero with the value 1 with the real version of Bernoulli's method.

[3] $p(z) = z^4 - 8z^3 + 40z^2 - 68z + 74$, see Essentials, Demonstration 3.2-3. Here the complex version of Bernoulli's method, i.e. with instruction 32 changed to GTO 33, is successful.

From $\hat{b} = 2 \operatorname{Re} z_1$ and $\hat{c} = |z_1|^2$ we find z_1 and $z_2 = \bar{z}_1$. If a polynomial of degree 4 is deflated with a quadratic factor we obtain a quadratic polynomial $q(z) := rz^2 + sz + t$. Clearly $r = 1$ and by comparing coefficients we get $-2.01... + s = -8$ and $9.96... \cdot t = 74$, from which we can compute s and t and the zeros $z_{3,4}$ of q and p.

[4] + [5] See Essentials, Demonstrations 3.2-4 and 3.2-5, where polynomials of degree 6 and 8 are considered. The program given in section 4 can only handle degrees $k \leq 4$. The results given in the Essentials have been obtained on a HP-67 computer.

Demonstrations 3.3-6 ÷ 3.3-8:

QUOTIENT DIFFERENCE ALGORITHM

1. Purpose

To compute <u>all</u> the zeros of a real polynomial of degree 4,

$$p(z) := a_0 z^4 + a_1 z^3 + a_2 z^2 + a_3 z + a_4 ,$$

<u>simultaneously.</u> (If the convergence is too slow, the obtained approximate values can be used as good starting values for a faster method for determining a single zero.)

2. Method

The progressive form of the qd algorithm, see Essentials, §3.3. For the present purpose it can be summarized as follows:

Let $a_0, a_1, a_2, a_3, a_4 \neq 0$; for $n = 0, 1, 2, \ldots$ we generate arrays of numbers

$$q_n^{(1)} \quad e_n^{(1)} \quad q_n^{(2)} \quad e_n^{(2)} \quad q_n^{(3)} \quad e_n^{(3)} \quad q_n^{(4)}$$

by the initial conditions

$$q_0^{(1)} := -\frac{a_1}{a_0}\,; \qquad q_0^{(k)} := 0\,, \qquad k = 2, 3, 4;$$

$$e_0^{(k)} := \frac{a_{k+1}}{a_k}\,, \qquad k = 1, 2, 3;$$

and the continuation rules

$$q_{n+1}^{(k)} = (e_n^{(k)} - e_n^{(k-1)}) + q_n^{(k)}\,,$$

$$e_{n+1}^{(k)} = \frac{q_{n+1}^{(k+1)}}{q_{n+1}^{(k)}}\, e_n^{(k)}\,,$$

where always

$$e_n^{(0)} := e_n^{(4)} := 0\,.$$

One such array of numbers consists of one q and the following e row of the qd scheme (see Essentials, Example 3.3-6). If the zeros z_k all have different absolute values, then the qd scheme exists and

$$\lim_{n\to\infty} q_n^{(k)} = z_k \qquad \text{and} \qquad \lim_{n\to\infty} e_n^{(k)} = 0$$

(convergence of the columns of the qd scheme, see Essentials, Theorems 3.3a and b). By means of auxiliary computation it is also possible to use these arrays to compute pairs of complex conjugate zeros (see Essentials, Theorem 3.3c). If, for instance, two q-columns, $q_n^{(1)}$ and $q_n^{(2)}$ say, show no convergence pattern and are flanked by e-columns which converge to (or are) zero, then z_1 and

z_2 can be found as zeros of the quadratic polynomial

$$z^2 - b_k z + c_k$$

where

$$b_k := \lim_{n\to\infty} (q_{n+1}^{(1)} + q_{n+1}^{(2)})$$

$$c_k := \lim_{n\to\infty} q_n^{(1)} \cdot q_{n+1}^{(1)} .$$

3. Flow Diagram

Straightforward. First initialize, then compute next array in the following order:

$$q_n^{(1)} \quad q_n^{(2)} \quad e_n^{(1)} \quad q_n^{(3)} \quad e_n^{(2)} \quad q_n^{(4)} \quad e_n^{(3)} \quad ;$$

the q-values are displayed.

4. Storage and Program

R_0	R_1	R_2	R_3	R_4	R_5	R_6	R_7
a_0	a_1	a_2	a_3	a_4			
n	$q^{(1)}$	$e^{(1)}$	$q^{(2)}$	$e^{(2)}$	$q^{(3)}$	$e^{(3)}$	$q^{(4)}$

$a_4 \in X$

	00	10	20	30	40
0		RCL 0	RCL 1	STO*2	RCL 6
1	RCL 3	STO÷0	R/S	RCL 6	STO-7
2	÷	CHS	RCL 4	RCL 4	RCL 7
3	STO 6	STO÷1	RCL 2	-	PAUSE
4	RCL 3	0	-	STO+5	RCL 5
5	RCL 2	STO 3	STO+3	RCL 5	÷
6	÷	STO 5	RCL 3	PAUSE	STO*6
7	STO 4	STO 7	PAUSE	RCL 3	1
8	RCL 1	→ RCL 2	RCL 1	÷	STO+0
9	STO÷2	STO+1	÷	STO*4	GTO 18

5. Operating Instructions

Load the program, then move the operating switch to RUN. Select mode of displaying numbers. Load the coefficients a_k of the given polynomial into R_k, k = 0, 1, 2, 3, 4 (a_0 is the coefficient of z^4 etc.!) and leave a_4 in the X-register (for reasons of programming space). Then press

PRGM

R/S

to start computation. The computer will then automatically generate the first two rows of the progressive form of the qd algorithm and stop by displaying $q_1^{(1)}$. For reasons of programming space $q_0^{(1)} = -\frac{a_1}{a_0}$ (see Essentials, Example 3.3-5) is never displayed and at this point already "overwritten" by $q_1^{(1)}$. But all the other quantities of the initial array can be inspected from the appropriate registers as indicated in section 4. This is the <u>only</u> place in the program, at which nondisplayed quantities may be recalled, because elsewhere, for reasons of programming space, the program is working with the quantities in the X-register. * Press

R/S

to continue the computation. The computer then will briefly display $q_1^{(2)}$, $q_1^{(3)}$ and $q_1^{(4)}$ and stop by displaying $q_2^{(1)}$; as above <u>only</u> now the other values of interest may be recalled. Now continue with * as above to get $q_3^{(1)}$ etc. Instead of the brief displays of $q_n^{(2)}$, $q_n^{(3)}$ and $q_n^{(4)}$ one can obtain stops by changing instructions 27, 36 and 43 to R/S; then R/S has to be pressed after each display to continue the computation. The e-values will always have to be recalled after the display of $q_n^{(1)}$.

6. Examples

[1] $p(z) := z^4 - 16z^3 + 72z^2 - 96z + 24$, see Essentials, Demonstration 3.3-6. Here $q_0^{(1)} = -\frac{-16}{1} = 16$.

[2] $p(z) := z^4 - 8z^3 + 39z^2 - 62z + 50$, see Essentials, Demonstration 3.3-7. After running the program, we discover that the polynomial must have two pairs of complex conjugate zeros. We hence compute (no special program, since the q-values have to be punched in anyway) the corresponding b and c values, to obtain the two quadratic factors of the polynomial as indicated in Demonstration 3.3-7.

[3] $p(z) := 81z^4 - 108z^3 + 24z + 20$, see Essentials, Demonstration 3.3-8. The qd scheme for $p^*(z^*) := 81z^{*4} + 216z^{*3} + 162z^{*2} + 24z^* + 17$ is (initial rows corresponding to $n = 0$ omitted):

n	$q_n^{(1)}$	$e_n^{(1)}$	$q_n^{(2)}$	$e_n^{(2)}$	$q_n^{(3)}$	$e_n^{(3)}$	$q_n^{(4)}$
1	-1.916667	0.235507	-0.601852	-0.137892	0.560185	-0.895661	-0.708333
2	-1.681159	0.136619	-0.975251	-0.027937	-0.197584	0.849168	0.187328
3	-1.544540	0.100819	-1.139807	0.016655	0.679521	-0.827074	-0.661840
4	-1.443721	0.085474	-1.223971	0.002234	-0.164208	0.832238	0.165234
5	-1.358247	0.082262	-1.307210	-0.001138	0.665795	-0.833749	-0.667004
6	-1.275986	0.089651	-1.390610	-0.000137	-0.166816	0.833393	0.166745
7	-1.186334	0.111874	-1.480398	0.000061	0.666714	-0.833312	-0.666648
8	-1.074460	0.165783	-1.592210	0.000006	-0.166660	0.833331	0.166663
9	-0.908678	0.320734	-1.757986	-0.000002	0.666665	-0.833334	-0.666667
10	-0.587944	1.133979	-2.078723	$-1.956743_{10}-7$	-0.166667	0.833333	0.166667

To obtain the two quadratic factors of the polynomial we form:

$b_n^{(0)}$	$c_n^{(0)}$	$b_n^{(2)}$	$c_n^{(2)}$
-2.518519	1.604939	-0.148148	0
-2.656410	1.869231	-0.010256	0.104938
-2.684347	1.916197	0.017681	0.130769
-2.667692	1.890472	0.001026	0.112280
-2.665457	1.887247	-0.001209	0.109527
-2.666596	1.888792	-0.000071	0.111018
-2.666732	1.888967	0.000066	0.111208
-2.666670	1.888893	0.000003	0.111117
-2.666664	1.888886	-0.000002	0.111107
-2.666667	1.888890	0.000000	0.111111
$\downarrow$	$\downarrow$	$\downarrow$	$\downarrow$
$-\frac{8}{3}$	$\frac{17}{9}$	0	$\frac{1}{9}$

Now we compute the zeros $z_1^* \div z_4^*$ and $z_1 \div z_4$ as indicated in Essentials, Demonstration 3.3-8.

Demonstrations 4.1-1, 4.1-2, 4.2-1 ÷ 4.2-3, 4.3-3, 4.4-2:

SIMULATION OF p-DIGIT DECIMAL ARITHMETIC

1. Purpose

To simulate a computer working with a mantissa of p decimal digits, $1 \leqq p \leqq 5$.

2. Method

Any given machine number x is rounded to a p-digit floating number $\tilde{x}$ in the following way:

(i) Check and store the sign ε of x and work with $|x|$.

(ii) Multiply $|x|$ with 10^q, where q - positive or negative - is chosen such that $|x| \cdot 10^q \in [10^{p-1}, 10^p]$; store $f := 10^{-q}$ ("correction factor").

(iii) Round the auxiliary number $|x| \cdot 10^q$ to a p-digit integer by forming

$$\mathrm{INT}(|x| \cdot 10^q + 0.5) \; .$$

(iv) Multiply this integer with $\varepsilon \cdot f$ to obtain $\tilde{x}$.

Now $\tilde{x}$ can be computed with; all the results of computations have to be rounded by the foregoing procedure.

3. <u>Flow Diagram</u>

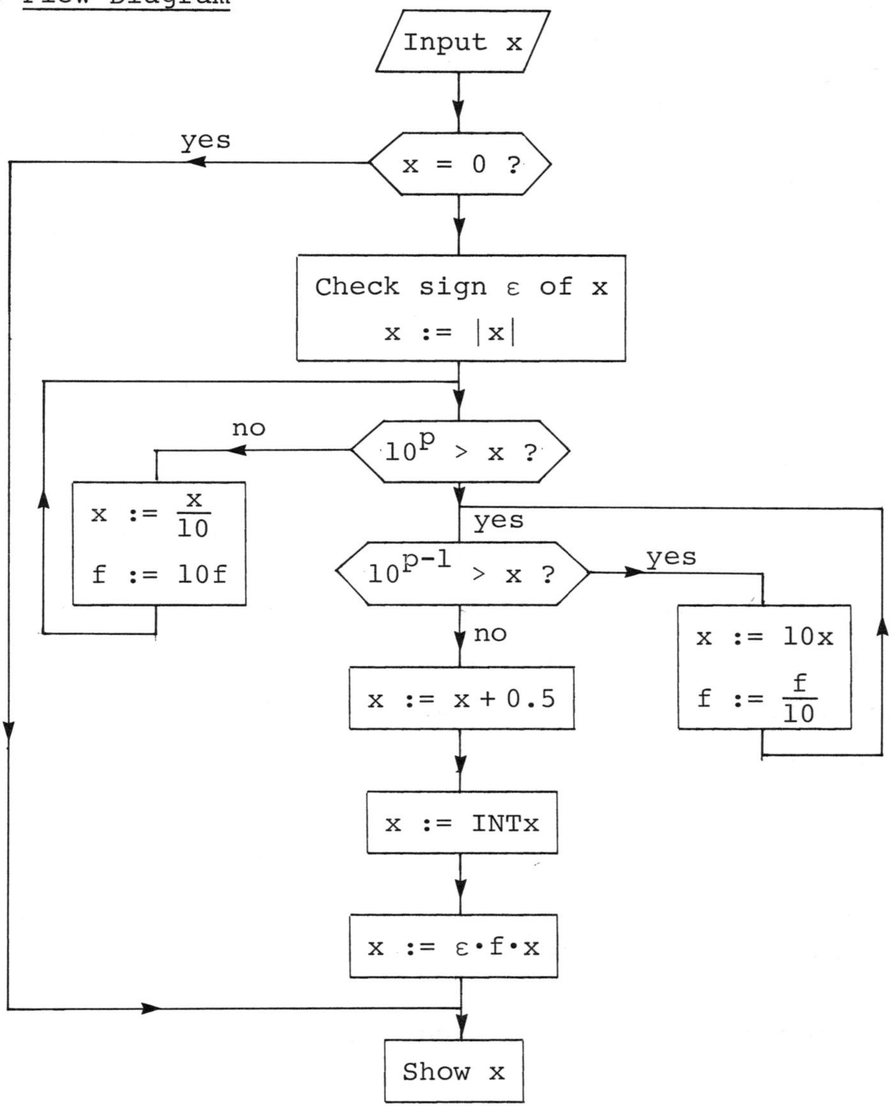

4. Storage and Program

R_0	R_1	R_2	R_3	R_4	R_5	R_6	R_7
10^p	$\varepsilon \cdot f$	10	$\tilde{y}$				

	00	10	20	30	40
0		ENTER	STO*1	•	RCL 2
1	1	ABS	÷	5	STO÷1
2	0	÷	GTO 15	+	*
3	STO 2	STO 1	→ $x \gtrless y$	INT	GTO 24
4	→ R/S	LAST X	→ RCL 0	RCL 1	
5	$x \gtrless y$	→ RCL 0	RCL 2	*	
6	STO 3	$x > y$	÷	→ RCL 3	
7	$x \gtrless y$	GTO 23	$x > y$	$x \gtrless y$	
8	$x = 0$	$x \gtrless y$	GTO 39	GTO 04	
9	GTO 36	RCL 2	$x \gtrless y$	→ $x \gtrless y$	

The program stores the previously rounded number $\tilde{y}$ into R_3 and preserves it (only) until the next number has been rounded, i.e. $\tilde{x}$ has been found. This enables the user to perform operations with $\tilde{x}$ and $\tilde{y}$ without storing them outside the computer.

5. Operating Instructions

Load program, move to RUN. Load 10^p into R_0; if $p = 4$ press

EEX
4

Choose mode of displaying numbers by pressing

SCI
6

[To show the correctly rounded p-digit numbers it would of course be sufficient to display just the $p < 6$ digits, but it is also interesting to see the zeros which have to appear after the rounding process.] Press

PRGM
R/S

The computer will stop right away, displaying 10. * Now load x into the X-register and press

R/S

to obtain $\tilde{x}$, the correctly rounded p-digit value of x. At this point the value shown previously can be found in R_3. Continue with * above. The registers $R_4 \div R_7$ can be used for additional storage (e.g. if one $\tilde{x}$ has to be preserved throughout several runs of the program).

6. Examples

[1] Computations of Demonstration 4.1-1, with $p = 4$, for instance to obtain the (3,3)-element of the first reduced tableau, which equals $0.2000 - 0.3333 * 0.3333$. The appearing product

is equal to 0.11108889 which is rounded to 0.1111. The difference then becomes 0.08890. See also Demonstrations 4.1-2, 4.2-1, 4.2-2, 4.2-3, 4.3-3, 4.4-2 for examples with $p = 4$.

2 Demonstration 4.3-2 is carried out with $p = 3$.

3 The foregoing program can also be used to simulate computations in "simple" precision with $p = 4$ versus "double" precision, i.e. full machine accuracy, see Demonstration 4.2-3.

Problem 4.2-7:

ITERATIVE REFINEMENT WITH "WRONG" L-R DECOMPOSITION

R_0	R_1	R_2	R_3	R_4	R_5	R_6	R_7
a	b	c	d	$x_1^{(0)}$	$x_2^{(0)}$	$-r_1$	$-r_2$
				$x_1^{(n)}$	$x_2^{(n)}$	w_1	w_2

	00	10	20	30	40
0		RCL 5	RCL 0	PAUSE	
1	1	2	STO÷6	RCL 3	
2	RCL 4	*	RCL 6	*	
3	2	+	RCL 1	STO-6	
4	*	CHS	*	RCL 6	
5	-	STO 7	STO-7	STO+4	
6	RCL 5	→ P	RCL 2	PAUSE	
7	-	PAUSE	STO÷7	GTO 01	
8	STO 6	PAUSE	RCL 7		
9	RCL 4	→ R	STO+5		

Applies iterative refinement to 2×2 system

$$2x_1 + x_2 = 1$$

$$x_1 + 2x_2 = 0 ,$$

with exact solution $x_1 = \frac{2}{3}$, $x_2 = -\frac{1}{3}$. If $(x_1^{(0)}, x_2^{(0)})$ is any trial solution, program shows length of corresponding residual vector $\underline{r}^{(0)}$ and computes corrections $\underline{c}^{(0)}$ from

$$\underline{L}\,\underline{w}^{(0)} = -\underline{r}^{(0)} ,$$

$$\underline{R}\,\underline{c}^{(0)} = \underline{w}^{(0)} ,$$

where

$$\underline{L} = \begin{pmatrix} a & 0 \\ b & c \end{pmatrix}, \qquad \underline{R} = \begin{pmatrix} 1 & d \\ 0 & 1 \end{pmatrix}$$

are alledged factors (possibly wrong) in L-R decomposition of matrix $\underline{A}$.

Similar program could be written for arbitrary 2×2 system $\underline{A}\,\underline{x} = \underline{v}$ where elements of $\underline{A}$ and of $\underline{v}$ are integers.

Demonstration 4.3-3:

STABLE LINEAR REGRESSION

1. Purpose

Given a sequence of data points (x_i, y_i), $i = 1, 2, \ldots, n$; to compute the coefficients of the linear regression function

$$y = ax + b \ .$$

We also wish to allow for the possibility of adding or removing a point from the set of data points.

2. Method

(See Essentials, §4.3, last section) Straightforward application of the method of least squares to this problem yields the formulas

$$\begin{aligned} a &= \frac{\sum x_i y_i - \frac{1}{n}\sum x_i \sum y_i}{\sum x_i^2 - \frac{1}{n}(\sum x_i)^2} \\ b &= \frac{1}{n}\sum y_i - a\frac{1}{n}\sum x_i \ , \end{aligned} \qquad \text{(I)}$$

in which all summations have to be extended from $i = 1$ to $i = n$. These formulas, in general, are unstable. However, if we introduce the running means

$$\mu := \frac{1}{n} \sum_{i=1}^{n} x_i \qquad \text{and} \qquad \nu := \frac{1}{n} \sum_{i=1}^{n} y_i$$

and shift the origin of our coordinate system into the center of mass of the given data points (and thereby leaving the slope a of the regression line unchanged), the formulas (I) become

$$a = \frac{\sum (x_i - \mu)(y_i - \nu)}{\sum (x_i - \mu)^2} ,$$

$$b = \nu - a\mu , \qquad \text{(II)}$$

and are stable. It is desirable to be able to compute the regression line dynamically, i.e. to add a new or remove an old data point without having to re-compute everything from scratch. To this end we define (now noting the dependence of the number n of data points)

$$q_n := \sum_{i=1}^{n} (x_i - \mu_n)^2 , \quad r_n := \sum_{i=1}^{n} (x_1 - \mu_n)(y_i - \nu_n)$$

and find recurrence relations

$$q_{n\pm1} =: Q(q_n) , \qquad r_{n\pm1} =: R(r_n) ,$$

(for details see Essentials, §4.3). In view of the implementation on the computer we again drop the subscripts and denote the new quantities obtained after adding or removing the point (x,y) with a prime: Let

$$n' := \begin{cases} n+1 , & \text{if the point is added} \\ n-1 , & \text{if the point is removed} , \end{cases}$$

then

$$q' = q + (n'-n)\frac{n'}{n}(x-\mu')^2$$

$$r' = r + (n'-n)\frac{n'}{n}(x-\mu')(y-\nu') .$$

If we define

$$s := \sum_{i=1}^{n} x_i , \qquad t := \sum_{i=1}^{n} y_i ,$$

we obtain

$$s' = s + (n'-n)x , \qquad t' = t + (n'-n)y$$

and

$$\mu' = \frac{1}{n'} s' , \qquad \nu' = \frac{1}{n'} t'$$

and finally

$$a = \frac{r'}{q'} , \qquad b = \nu' - a\mu' . \tag{II'}$$

3. Flow Diagram

In order to demonstrate the difference between the unstable and stable versions, we provide programs based on both sets of formulas, (I) and (II'). In order to add or remove a data point, the key $\Sigma+$ or $\Sigma-$ has to be pressed when running either program thereby updating n and all the sums automatically. (The sums which are not needed to implement (I) or (II') are either neglected or overwritten with temporary data; see also HP-33E manual, "statistical functions" for definition and location of sums.) To compute means, the instruction $\bar{x}$ is used in both programs.

(I) The flow diagram is straightforward and hence omitted here. If only one data point has been read in, the calculator first "asks" for the next data point, before computing a and b (a would not be defined).

(II)

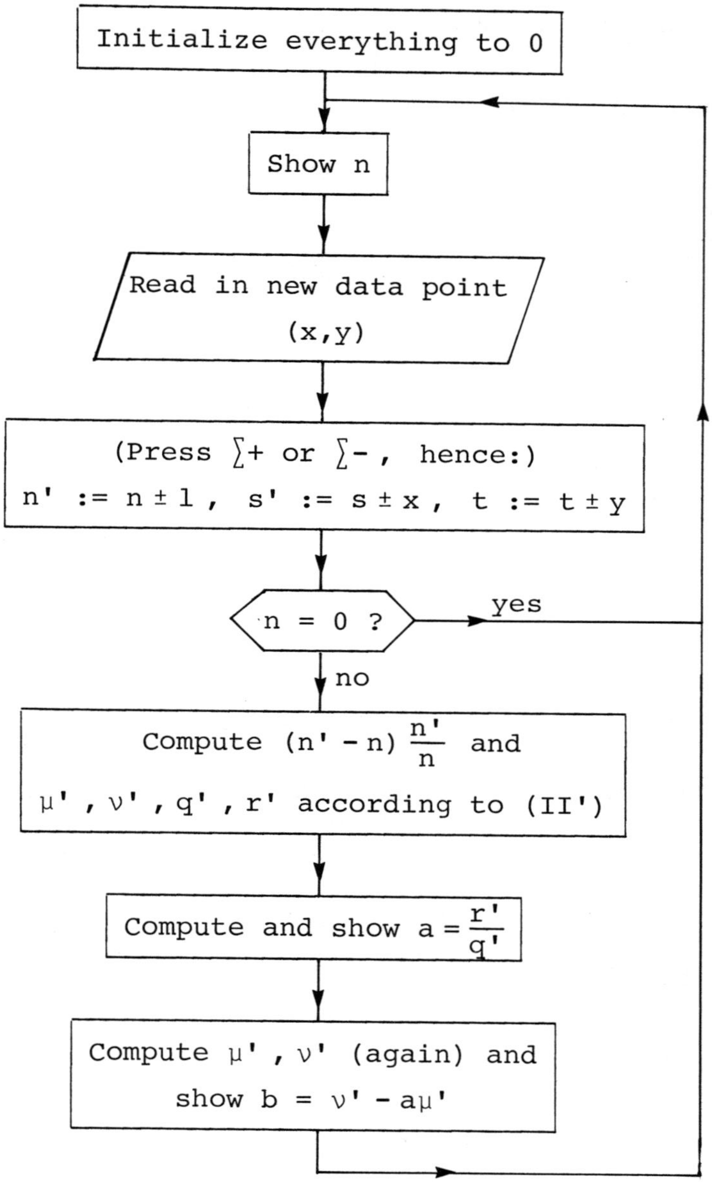
Initialize everything to 0
Show n
Read in new data point (x,y)
(Press ∑+ or ∑-, hence:)
n' := n ± 1, s' := s ± x, t := t ± y
n = 0 ?
yes
no
Compute (n' - n) n'/n and
μ', ν', q', r' according to (II')
Compute and show a = r'/q'
Compute μ', ν' (again) and
show b = ν' - aμ'

4. Storage and Program

(I)	R_0	R_1	R_2	R_3	R_4	R_5	R_6	R_7
	a		n'	$s=\sum x$	$\sum x^2$	$t=\sum y$	$\sum y^2$	$\sum xy$

	00	10	20	30	40
0	⊠	RCL 7	RCL 2	R/S	
1	REG	RCL 3	÷	GTO 02	
2	→ RCL 2	RCL 5	−		
3	FIX 0	*	÷		
4	R/S	RCL 2	R/S		
5	FIX 8	÷	STO 0		
6	R ↓	−	$\bar{x}$		
7	R ↓	RCL 4	RCL 0		
8	x = 0	RCL 3	*		
9	GTO 02	x^2	−		

(II')	R_0	R_1	R_2	R_3	R_4	R_5	R_6	R_7
	q	r	n'	s	x	t	y	$(n'-n)\frac{n'}{n}$

	00	10	20	30	40
0	⊠	STO 4	STO 7	x ≷ y	R/S
1	REG	R ↓	$\bar{x}$	RCL 6	STO 7
2	→ RCL 2	x ≷ y	RCL 4	−	$\bar{x}$
3	FIX 0	x = 0	−	*	RCL 7
4	R/S	GTO 02	ENTER	RCL 7	*
5	FIX 8	−	x^2	*	−
6	x ≷ y	LAST X	RCL 7	STO+1	R/S
7	STO 6	÷	*	RCL 1	GTO 02
8	R ↓	RCL 2	STO+0	RCL 0	
9	LAST X	*	R ↓	÷	

5. <u>Operating Instructions</u>

They are the same for both programs. Load program (I) or (II') and move to RUN. Press

PRGM
R/S

* On the display of n. (n = number of data points already taken into account), load the data point (x,y) such that $x \in X$ and $y \in Y$ (thus y is to be loaded first). If the data point is <u>added,</u> press

Σ+
R/S

if the data point is <u>removed,</u> press

Σ-
R/S

Unless n was zero, when the calculator goes back to * right away, the updated value of a will be displayed, and after pressing

R/S

the updated value of b will be shown. Press

R/S

to get back to *, etc.

6. Example

Let $x_k = \pi,\ e,\ e^{-\pi} + 10^{2m},\ m = 0, 1, 2, 3,$

$$y_k = \sqrt{2}\, x_k + 1/3,$$

see Essentials, Demonstration 4.3-3.

Problem 5.1-3:

BERNSTEIN POLYNOMIALS

R_0	R_1	R_2	R_3	R_4	R_5	R_6	R_7
$\boxed{n}$	x	x_k	k	c_k	$\sum$		
n-k							
x∈X							

	00	10	20	30	40
0		÷	x_k ∈ X	RCL 4	RCL 0
1	STO 1	STO 2	and	*	-
2	RCL 0	Pro-	$x_k \in R_2$	STO+5	STO÷4
3	STO 3	gram		RCL 3	1
4	y^x	to	and	x = 0	RCL 1
5	STO 4	com-	putting	GTO 49	1/x
6	0	pute	$f(x_k)$	STO*4	-
7	STO 5	$f(x_k)$,		1	STO*4
8	RCL 3	as-	into	-	GTO 08
9	RCL 0	suming	X	STO 3	→ RCL 5

Evaluates the n-th Bernstein polynomial associated with f,

$$p_n(x) := \sum_{k=0}^{n} f(\frac{k}{n})\, x^k\, (1 - x)^{n-k} ,$$

for $x \in (0,1]$.

Demonstrations 5.4-1, 5.4-2, 5.4-3:

LAGRANGIAN INTERPOLATION: EQUIDISTANT KNOTS

1. Purpose

To evaluate, for any $n \geqq 0$ the polynomial p that interpolates a given function f at $n+1$ equidistant interpolating points $x_k := x_0 + k \cdot h$, $k = 0, 1, \ldots, n$, where h is a positive constant.

2. Method

Lagrangian interpolation. Evaluation by means of the barycentric formula, see Essentials, §5.4,

$$p(x) = \frac{\sum_{k=0}^{n} w_k^* \frac{f_k}{x - x_k}}{\sum_{k=0}^{n} w_k^* \frac{1}{x - x_k}},$$

where $f_k := f(x_k)$ and the w_k^* are the modified (such that $w_0^* = 1$) weights, which in the case of equidistant knots equal

$$w_k^* = (-1)^k \binom{n}{k} = (-1)^k \frac{n!}{k!(n-k)!},$$

see Essentials, §5.4. The weights are computed recursively by

$$w_0^* = 1 \, , \qquad w_{k+1}^* = \frac{n-k}{-1-k} \, w_k^* \, .$$

3. Flow Diagram

See Essentials, Fig. 5.4a. s and t denote numerator and denominator of the barycentric formula. To avoid division by 0 if $x = x_k$, the program checks whether $x - x_k = 0$ and if so replaces 0 by 10^{-30}; this generally will produce the correct display of $p(x_k) = f_k$, see Essentials, § 5.4.

4. Storage and Program

R_0	R_1	R_2	R_3	R_4	R_5	R_6	R_7
w*	-k	t	s	h	n	x_0	x
					n-k	x_k	

	00	10	20	30	40
0		-	GSB 37	RCL 1	
1	1	x ≠ 0	*	STO÷0	
2	STO 0	GTO 18	STO+3	RCL 4	
3	CLX	CLX	RCL 5	STO+6	
4	STO 1	EEX	x = 0	GTO 07	
5	STO 2	CHS	GTO 35	$\bar{x}$	
6	STO 3	3	STO*0	R/S	
7	→ RCL 0	0	1	→	
8	RCL 7	→ ÷	STO-5		
9	RCL 6	STO+2	STO-1		

Instructions 37 ÷ 49 are available for the subroutine computing f(x). This subroutine should assume x in R_6 and leave the computed value f(x) in the X-register. The subroutine must not change the content of the Y-register from the main program. Its last instruction should be RTN. Instruction 35 not only produces the desired quotient $\frac{s}{t}$ but also an unnecessary and "harmless" quotient; therefore we prefer the use of $\bar{x}$ to the longer version

```
RCL 3
RCL 2
÷
```

5. Operating Instructions

Load program, including subroutine to compute f(x). Move to RUN and select mode of displaying numbers. Load data as follows

h (= distance between knots)
into R_4

n (= degree of interpolating polynomial
= number of knots minus one)
into R_5

x_0 (= leftmost knot)
into R_6

x (= point at which polynomial is evaluated) into R_7

Pressing

PRGM
R/S

starts the computation. The calculator stops by displaying p(x). If the calculation is to be repeated for a different x, then (for reasons of storage space) the values of n and x_0 must be re-stored in the R_5 and R_6 registers (h need not be re-stored), and the new x must be loaded into R_7. Pressing

R/S

will yield new value of p(x).

The program may also be used to interpolate a function that is not generated internally. In this case the subroutine providing the values of the function should look like this:

```
37    RCL 6
38    R/S
39    x ≷ y
40    R ↓
41    RTN
```

(Instructions 39 and 40 are necessary to re-store the Y-register of the main program.) After loading the data and pressing

PRGM
R/S

as above, the calculator then will halt and display x_0. One then should read in $f(x_0)$ into the X-register and press

R/S

The calculator will stop and display x_1. One then should read in $f(x_1)$. This process is continued until all $f(x_k)$ have been read in. After pressing

R/S

one last time the calculator displays p(x).

6. <u>Examples</u>

[1] Cubic interpolation of f(x) = sin x (x given in degrees) in the interval [44,46], using the interpolating points 42, 44, 46, 48, see Essentials, §5.4, Demonstration 5.4-1. The subroutine for f reads

```
37   RCL 6
38   DEG
39   SIN
40   RTN
```

and the input data are $h = 2$, $n = 3$, $x_0 = 42$ with $x = 44.2, 44.4, 44.6, \ldots, 45.8$.

[2] Interpolation of $f(x) = x$, at $x = 0.5$, by a polynomial of degree n, using $x_k = k$, $k = 0, 1, \ldots, n$, see Essentials, §5.4, Demonstration 5.4-2. The subroutine is

```
37   RCL 6
38   RTN
```

and the input data are $h = 1$, $n = 2^{\ell}$ $(\ell = 0,1,\ldots,6)$, $x_0 = 0$, $x = 0.5$.

[3] To compute $\sin(\pi/4)$ by interpolating the function $f(x) = \sin x$ (x given in radians) by a polynomial of degree $n = 2\ell + 1$, $\ell = 0, 1, 2, \ldots, 63$ using the interpolating points

$$\frac{\pi}{4} \pm \left(\frac{\pi}{4} + m \cdot \frac{\pi}{2}\right) ,$$

$m = 0, 1, 2, \ldots, \ell$, see Essentials, §5.4, Demonstration 5.4-3. We use the subroutine

```
37   RCL 6
38   RAD
39   SIN
40   RTN
```

The input data are

$h = \pi/2$, $n = 1$, $x_0 = 0$, $x = \pi/4$
" $n = 3$, $x_0 = -\pi/2$, "
" $n = 5$, $x_0 = -\pi$, "
" $n = 7$, $x_0 = -3\pi/2$, "
" $n = 9$, $x_0 = -2\pi$, "
" $n = 11$, $x_0 = -5\pi/2$, "

. . .

" $n = 127$, $x_0 = -63\pi/2$, "

4 Here we demonstrate the use of the program for manual interpolation. Given the following values of the Bessel function J_0:

	x	$J_0(x)$
	2.2	0.110362267
	2.3	0.055539784
	2.4	0.002507683
→	2.5	-0.048383776
	2.6	-0.096804954
	2.7	-0.142449370

We interpolate the value of J_0 at $x = 2.404825558$, the first zero of J_0. We work with $n = 1, 3, 5$, always using the points x_k closest to x. Results:

n	p(x)
1	0.000051886
3	0.000000085
5	$-2.6 * 10^{-10}$

Problem 5.4-2:

EXTRAPOLATION TO $x = 0$ FROM VALUES AT GEOMETRIC PROGRESSION

R_0	R_1	R_2	R_3	R_4	R_5	R_6	R_7
n	q	t	s	q^k	w		
n-k							

	00	10	20	30	40
0	⨯	to	and	STO+2	1
1	0	com-		*	STO-0
2	STO 2	pute	$q^k \in X$,	STO+3	-
3	STO 3		and	RCL 1	STO*5
4	1	$f(q^k)$,	putting	STO*4	RCL 4
5	STO 4			RCL 0	1
6	STO 5	as-	$f(q^k)$	x = 0	-
7	→ RCL 4	suming		GTO 49	STO÷5
8	Pro-		into X	CHS	GTO 07
9	gram	$q^k \in R_4$	RCL 5	y^x	→ $\bar{x}$

Evaluates $p(0)$ where $p(x)$ is polynomial of degree n interpolating $f(x)$ at $x = 1, q, \ldots, q^n$. $p(0)$ is evaluated in barycentric form,

$$p(0) = \frac{\sum_{k=0}^{n} w_k^* f(q^k)}{\sum_{k=0}^{n} w_k^*} ,$$

where $w_0^* = 1$, $w^*_k = -\frac{1 - q^{k-n-1}}{1 - q^k} w_{k-1}^*$.

Demonstrations 5.5-1, 5.5-2:

LAGRANGIAN INTERPOLATION: CHEBYSHEV KNOTS

1. Purpose

To evaluate, for any $n > 0$, the polynomial p that interpolates a given function f, defined on the interval $[-1,+1]$ at the $n+1$ "Chebyshev knots"

$$x_k := \cos \phi_k , \qquad \phi_k := (k+\tfrac{1}{2}) \frac{\pi}{n+1} , \qquad k = 0,1,\ldots,n.$$

2. Method

As in the previous program ("Lagrangian Interpolation: Equidistant knots"), the interpolating polynomial is evaluated by means of the barycentric formula. For the modified weights we obtain

$$w_k^* = (-1)^k \sin \phi_k ,$$

see Essentials, §5.5. To save programming space, these weights are computed as

$$w_k^* = (-1)^k \sin\phi_k = \cos k\pi \cdot \sin\phi_k + 0$$

$$= \sin(\phi_k + k\pi)$$

$$= \sin(\phi_k + (2k+1)\frac{\pi}{2} - \frac{\pi}{2})$$

$$= \sin((n+2)\phi_k - \frac{\pi}{2})$$

$$= \cos((n+2)\phi_k) .$$

Thus the weights are not computed recursively and therefore do not need to be stored.

3. Flow Diagram

We denote the numerator and denominator in the barycentric formula by s and t respectively. We note that the increasing sequence of the angles ϕ_k starts with $\phi_0 = \frac{\pi}{2(n+1)}$ and ends with $\phi_n = \frac{2n+1}{2n+2}\pi < \pi$; the increment is always $\Delta\phi = \frac{\pi}{n+1}$.

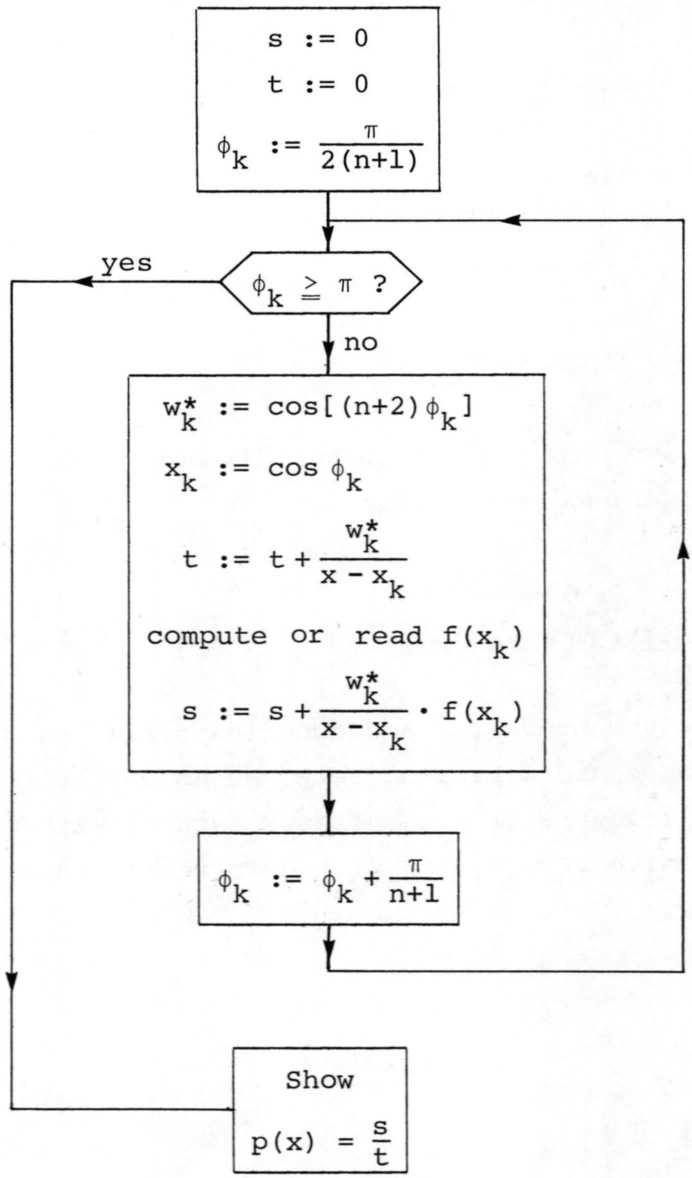

$s := 0$
$t := 0$
$\phi_k := \frac{\pi}{2(n+1)}$
yes
$\phi_k \geq \pi$?
no
$w_k^* := \cos[(n+2)\phi_k]$
$x_k := \cos \phi_k$
$t := t + \frac{w_k^*}{x - x_k}$
compute or read $f(x_k)$
$s := s + \frac{w_k^*}{x - x_k} \cdot f(x_k)$
$\phi_k := \phi_k + \frac{\pi}{n+1}$
Show
$p(x) = \frac{s}{t}$

4. Storage and Program

R_0	R_1	R_2	R_3	R_4	R_5	R_6	R_7
	ϕ	t	s	$\Delta\phi = \frac{\pi}{n+1}$	$\boxed{n}$ not destroyed	x_k	$\boxed{x}$ not destroyed

	00	10	20	30	40
0		STO 4	+	GSB 39	
1	RAD	2	*	*	
2	CLX	÷	cos	STO+3	
3	STO 3	STO 1	RCL 7	RCL 4	
4	STO 2	→ π	RCL 1	STO+1	
5	π	x ≤ y	cos	RCL 1	
6	RCL 5	GTO 37	STO 6	GTO 14	
7	1	RCL 1	-	→ $\bar{x}$	
8	+	RCL 5	÷	GTO 00	
9	÷	2	STO+2	→	

Instructions 39 ÷ 49 are available for the subroutine computing f(x). This subroutine should assume x in R_6 and leave the computed value f(x) in the X-register, and at the same time it must not change the content of the Y-register from the main program. Its last instruction should be RTN. Instruction 37 not only computes the desired quotient $\frac{s}{t}$ but also $\frac{n}{t}$ which, being in the Y-register at the end of the computation, does not do any harm; but the use of $\bar{x}$ saves us two programming steps.

5. Operating Instructions

Load program, including subroutine to compute f(x). Move to RUN and select mode of displaying numbers. Load data as follows

n (= degree of interpolating polynomial
= number of knots minus one)
into R_5

x (= point at which polynomial is evaluated)
into R_7

Pressing

PRGM
R/S

starts the computation. The calculator stops by displaying p(x). If the calculation is to be repeated for a different x, load new value of x into R_7 and press

R/S

The calculator will stop by displaying p(x), etc.

The program may also be used to interpolate a function that is not generated internally. In this case the subroutine providing the values of the function should look like this:

```
RCL 6
R/S
x ≷ y
R ↓
RTN
```

After loading the data and pressing

PRGM
R/S

the calculator will halt and display x_0. Read in $f(x_0)$ into the X-register and press

R/S

The calculator will stop displaying x_1. Read in $f(x_1)$ etc. until all $f(x_k)$ have been read in. After pressing

R/S

one last time the calculator displays p(x).

<u>Caution:</u> If accidentally x equals one of the knots x_k, an error halt results, because the barycentric formula then requires a division by zero. However, unless $f(x_k) = 0$ the correct value of p(x) will generally be obtained by loading on display "error" a very small number, performing manually the following steps:

```
R ↓         *)
CLX
EEX
CHS
3
0
R/S
```

[*) : To erase "error" any key could be pressed, but the R ↓ instruction can be thought of as bringing the unwanted denominator 0 on display.)

6. Examples and Timing

[1] $f(x) := x$, see Essentials, Demonstration 5.5-1. The subroutine for f simply is:

```
39   RCL 6
40   RTN
```

[2] $f(x) := |x|$, see Essentials, Demonstration 5.5-2; subroutine for f:

```
39   RCL 6
40   ABS
41   RTN
```

3 $f(x) := \arctan(x)$. Here the subroutine for f is:

```
39   RCL 6
40   5
41   *
42   TAN^-1
43   RTN
```

(The instruction RAD is already contained in the main program.)

n	x = 0.2	x = 1.0
2	0.31034546	1.55172728
4	0.49728439	1.31741228
8	0.69954553	1.36229895
16	0.78815414	1.37239249
32	0.78532124	1.37338140
[∞	$\frac{\pi}{4} = 0.78539816$	arctan5 = 1.37340077]

Problem 5.5-3:

SALZER'S FORMULA

R_0	R_1	R_2	R_3	R_4	R_5	R_6	R_7
$\boxed{n}$	x	D	N	k	$(-1)^k$	$\cos\frac{k\pi}{n}$	

$x \in X$

	00	10	20	30	40
0	⊠	RCL 4	$f(x_k)$	RCL 6	*
1	RAD	→ π	into	−	STO+3
2	STO 1	*	X	÷	1
3	RCL 0	RCL 0	as-	1	CHS
4	STO 4	÷	suming	RCL 6	STO*5
5	0	cos	$x_k \in X$	ABS	STO+4
6	STO 2	STO 6	and	+	RCL 4
7	STO 3	Pro-	$x_k \in R_6$	INT	x > y
8	1	gram	RCL 5	÷	GTO 11
9	STO 5	placing	RCL 1	STO+2	$\bar{x}$

Evaluates polynomial p(x) interpolating f(x) at $x_k := \cos\frac{k\pi}{n}$, $k = 0,1,\ldots,n$, in barycentric form due to Salzer,

$$p(x) = \frac{\sum_{k=0}^{n} (-1)^k \varepsilon_k \frac{f(x_k)}{x - x_k}}{\sum_{k=0}^{n} (-1)^k \varepsilon_k \frac{1}{x - x_k}},$$

where $\varepsilon_0 = \varepsilon_n = \frac{1}{2}$, all other $\varepsilon_k = 1$.

Demonstration 5.6-1:

2-POINT HERMITE INTERPOLATION, DATA PRESERVED

1. Purpose

Given a differentiable function f defined on an interval I containing the two points x_0 and x_1. Let $f_k := f(x_k)$, $f'_k := f'(x_k)$, $k = 0, 1$. To evaluate the polynomial p(x) of degree $\leqq 3$, which interpolates f and has the same slope as f at the two given points, i.e. such that

$$p(x_k) = f_k \ , \qquad p'(x_k) = f'_k \ , \qquad k = 0, 1.$$

This is a special case of Hermite interpolation in two ways:

(i) only the <u>first derivative</u> (instead of m_k derivatives at the point x_k) has to agree at <u>each</u> point,

(ii) there are only <u>two</u> (instead of n) <u>interpolating points.</u>

Exactly one such polynomial exists. For details see Essentials, §5.6. The input data x_k, f_k, f'_k, $k = 0, 1$, should be preserved.

2. Method

An application of a barycentric formula for the evaluation of the Hermite interpolating polynomial to our special case yields

$$p(x) = \frac{1}{\Delta^2}\Big\{(x-x_1)^2 f_0 + (x-x_0)^2 f_1$$

$$+ (x-x_0)(x-x_1)\big[(x-x_1)(f_0'+\tfrac{2}{\Delta}f_0)+(x-x_0)(f_1'-\tfrac{2}{\Delta}f_1)\big]\Big\},$$

where $\Delta := x_1 - x_0$, see Essentials, §5.6.

3. Flow Diagram

Basically straightforward; since there are only two storage registers available, the stack had to be used extensively, which sometimes obliterates the clarity of the program.

4. Storage and Program

R_0	R_1	R_2	R_3	R_4	R_5	R_6	R_7
x_0	x_1	f_0	f_1	f'_0	f'_1	$\frac{x-x_0}{\Delta}$	$\frac{x-x_1}{\Delta}$

	00	10	20	30	40
0		RCL 4	RCL 3	+	x^2
1	ENTER	RCL 1	2	RCL 7	*
2	ENTER	RCL 0	*	*	+
3	RCL 0	-	-	+	RCL 3
4	-	STO÷6	RCL 6	RCL 7	RCL 6
5	STO 6	STO÷7	*	*	x^2
6	R ↓	*	x ≷ y	RCL 6	*
7	RCL 1	LAST X	RCL 2	*	+
8	-	RCL 5	2	RCL 2	GTO 00
9	STO 7	*	*	RCL 7	

5. Operating Instructions

Load program. Move to RUN. Select mode of displaying numbers, e.g. by pressing

FIX 9

Load data as follows

x_0 into R_0
x_1 into R_1
f_0 into R_2
f_1 into R_3

f'_0 into R_4

f'_1 into R_5

Now load the value of x, at which p has to be evaluated into the X-register and press

PRGM

R/S

The calculator will stop and show p(x). If the value of the polynomial for a different x is desired, load the new x into the X-register and press again

PRGM

R/S

etc. (The input data do not need to be re-loaded.)

6. <u>Example</u>

$f(x) := \sin x$ (with x measured in degrees), $x_0 := 44$, $x_1 := 46$, see Essentials, Demonstration 5.6-1. After pressing DEG we obtain the input data

$$f_0 = \sin 44 = 0.694658371$$

$$f_1 = \sin 46 = 0.719339800$$

$$f'_0 = \frac{\pi}{180} \cos 44 = 0.012554848$$

$$f'_1 = \frac{\pi}{180} \cos 46 = 0.012124076$$

Demonstration 5.7-1:

SOLVING EQUATION BY INVERSE 2-POINT HERMITE INTERPOLATION

1. Purpose

To solve the scalar non-linear equation

$$f(x) = 0$$

by means of interpolating the inverse function $f^{[-1]}$ and evaluating the resulting interpolating polynomial $p(y)$ at $y = 0$.

2. Method

Iterated inverse 2-point Hermite interpolation, see Essentials, §5.7: Choose starting values x_0 and x_1, such that $y_0 := f(x_0)$ and $y_1 := f(x_1)$ have alternating signs. Evaluate y_0, y_1, f'_0, f'_1. Compute x_2 from

$$x_2 := \frac{1}{(y_1-y_0)^2}\left\{x_0y_1^2+x_1y_0^2-y_0y_1\left[\left(\frac{1}{f'_0}+\frac{2x_0}{y_1-y_0}\right)y_1+\left(\frac{1}{f'_1}-\frac{2x_1}{y_1-y_0}\right)y_0\right]\right\}.$$

If $y_2 := f(x_2) = 0$ then x_2 is the desired solution. Otherwise substitute $x_0 := x_1$, $x_1 := x_2$ and repeat the algorithm.

3. <u>Flow Diagram</u>

Straightforward.

4. <u>Storage and Program</u>

R_0	R_1	R_2	R_3	R_4	R_5	R_6	R_7
x_0	x_1	y_0	y_1	$\frac{1}{f'(x_0)}$	$\frac{1}{f'(x_1)}$	y_0-y_1	*

$x_1 \in X$

	00	10	20	30	40
0		x_1	RCL 2	RCL 5	x^2
1	Put	in	RCL 3	STO 4	÷
2	y_1	X	-	RCL 7	STO-7
3	into	x_1	STO 6	-	RCL 3
4	R_3	in	÷	RCL 2	STO 2
5	$\frac{1}{f'(x_1)}$	R_1	STO 7	*	STO*7
6		RCL 0	RCL 4	-	RCL 7
7	into	RCL 1	-	RCL 2	PAUSE
8	R_5	STO 0	RCL 3	*	STO-1
9	assuming	-	*	RCL 6	RCL 1

Note: A more elegant program could be written by using subroutines for the computation of y_1 and $1/f'(x_1)$. But this would require a few more programming places (e.g. GSB .., RTN, etc.), which we prefer to save for more complicated functions f.

5. Operating Instructions

Load program, including program to compute y_1 and $1/f'(x_1)$. Move to RUN. Load starting values as follows:

x_0	into	R_0
x_1	into	R_1
y_0	into	R_2
$1/f'(x_0)$	into	R_4

<u>and put</u> x_1 <u>into the</u> X-<u>register</u>. Choose mode of displaying numbers. Now press

PRGM
R/S

The computer will briefly display $x_1 - x_2$ and then stop at the display of x_2. For a new iteration press

R/S

The convergence has to be eye-checked. Since the computation of x_2 requires a division through $y_0 - y_1$, the algorithm must be stopped when $y_0 - y_1 = 0$. If the zero of f by then has not been found, the algorithm has to be restarted with new starting values.

6. Example

$f(x) := x^2 - 4$, $x_0 := 1$, $x_1 := 10$, see Essentials, Demonstration 5.7-1. The initial program for this function is:

00		10	NOP
01	x^2	11	NOP
02	4	12	NOP
03	-	13	NOP
04	STO 3	14	NOP
05	RCL 1	15	NOP
06	2		
07	*		
08	1/x		
09	STO 5		

Input data: $y_0 = -3$, $1/f'(x_0) = 1/2$.

Demonstration 5.9-1:

SPLINE INTERPOLATION

1. Purpose

Given a function f which is defined at least at the $n+1$ equidistant points $x_k := x_0 + k \cdot h$, $k = 1, 2, \ldots, n$, $h > 0$. To construct its spline interpolant g, i.e. the uniquely determined (see Essentials, Theorem 5.8) function g with the properties (i)-(iv) of Essentials, §5.8.

2. Method

Spline interpolation as described in Essentials, §5.8 (theory) and §5.9 (algorithm). g in each interval $[x_i, x_{i+1}]$ is represented by a cubic polynomial $q_i(t)$, the evaluation of which requires the knowledge of the derivatives s_i of g at the interpolating points x_i and x_{i+1}. We recall that the vector $\underline{s}$ of the unknown $h \cdot s_i$ can be determined as solution of $\underline{A}\,\underline{s} = \underline{b}$, where $\underline{A}$ is tridiagonal (and independant of f) and $\underline{b}$ is the input vector

$$\underline{b} := \begin{pmatrix} b_0 \\ b_1 \\ b_2 \\ \vdots \\ b_{n-1} \\ b_n \end{pmatrix} = \begin{pmatrix} 3f_1 - 3f_0 \\ 3f_2 - 3f_0 \\ 3f_3 - 3f_1 \\ \vdots \\ 3f_n - 3f_{n-2} \\ 3f_n - 3f_{n-1} \end{pmatrix}, \qquad f_i := f(x_i) .$$

The algorithm proceeds in three phases.

<u>Phases I and II:</u> Find the L-R decomposition of $\underline{A}$. All elements $\neq 0$ or 1 of $\underline{L}$ and $\underline{R}$ can be expressed in terms of the r_k, $k = 0, 1, \ldots, n-1$, i.e. the upper main diagonal of $\underline{R}$. The following formulas result:

$$r_0 = \frac{1}{2}, \qquad r_i = \frac{1}{4 - r_{i-1}}, \qquad i = 1, 2, \ldots, n-1,$$

and on the HP-33E with a 9-digit mantissa we may put

$$r_i := 2 - \sqrt{3} = 0.267949192, \qquad i \geqq 9 .$$

I: Solution of $\underline{L}\,\underline{y} = \underline{b}$:

$$y_0 = r_0 b_0 ,$$

$$y_i = r_i (b_i - y_{i-1}), \qquad i = 1, 2, \ldots, n-1,$$

$$y_n = \frac{b_n - y_{n-1}}{2 - r_{n-1}} .$$

II: Solution of $\underline{R}\,\underline{s} = \underline{y}$:

$$hs_n = y_n$$

$$hs_i = y_i - r_i(hs_{i+1}) \,, \qquad i = n-1, n-2, \ldots, 0.$$

Phase III: Evaluation of g at x in interval containing all x_k :

1) Determine interval $[x_i, x_{i+1}]$, in which x lies, i.e. calculate integral part

$$i := \left[\frac{x - x_0}{h}\right] .$$

2) Calculate $t := \dfrac{x - x_i}{h}$,

$$q(t) = (1-t)^2 f_i + t^2 f_{i+1}$$
$$+ t(1-t)[(1-t)(2f_i + hs_i) + t(2f_{i+1} - hs_{i+1})].$$

3. Flow Diagram

For reasons of programming space each phase of the algorithm had to be implemented by a separate program.

Phase I: See Essentials, Figure 5.9a.

Phase II: Straightforward.

Phase III: On a larger computer, the flow diagram would be completely straightforward, i.e. first read x_0, h, x, then compute i, t and $1-t$, then read f_i, f_{i+1}, hs_i, hs_{i+1} and finally compute q(t). Program III given below, for reasons of storage space, differs from this, inasmuch as it first reads f_i and hs_i and computes the terms involving these quantities and only then reads f_{i+1} and hs_{i+1} to finish the computation of q(t).

4. Storage and Program

Phase I:

R_0	R_1	R_2	R_3	R_4	R_5	R_6	R_7
[n]	i	r_{i-1}	$-y_{i-1}$				

	00	10	20	30	40
0	╳	STO+3	STO 2	RCL 2	
1	0	RCL 1	RCL 3	-	
2	STO 1	RCL 0	*	STO÷3	
3	STO 3	-	R/S	RCL 3	
4	2	x = 0	CHS	R/S	
5	STO 2	GTO 29	STO 3	0	
6	→ RCL 1	4	1	FIX 2	
7	FIX 0	RCL 2	STO+1	GTO 00	
8	R/S	-	GTO 06		
9	FIX 8	1/x	→ 2		

Phase II:

R_0	R_1	R_2	R_3	R_4	R_5	R_6	R_7
n		r_i		i	y_{i+1}		

	00	10	20	30	40
0		x < 0	9	-	→ STO*5
1	RCL 0	GTO 00	1	1/x	RCL 0
2	FIX 0	x ≦ y	9	STO 2	FIX 0
3	R/S	GTO 25	2	1	R/S
4	FIX 8	•	GTO 40	STO-4	FIX 8
5	STO 5	2	→ STO 4	RCL 4	RCL 5
6	→ 1	6	2	x < 0	-
7	STO-0	7	STO 2	GTO 39	R/S
8	9	9	→ 4	GTO 28	STO 5
9	RCL 0	4	RCL 2	→ RCL 2	GTO 06

Phase III:

R_0	R_1	R_2	R_3	R_4	R_5	R_6	R_7
x_0	h	t	1-t	k	*	*	*

$x \in X$

	00	10	20	30	40
0		STO 2	RCL 3	STO 6	STO+5
1	RCL 0	-	STO*5	x ≷ y	RCL 2
2	-	STO 3	x^2	STO+6	STO*5
3	RCL 1	RCL 4	*	STO+6	RCL 3
4	÷	FIX 0	STO 7	RCL 2	STO*5
5	INT	R/S	1	STO*6	RCL 5
6	STO 4	STO 5	STO+4	x^2	RCL 7
7	1	x ≷ y	RCL 4	*	+
8	LAST X	STO+5	R/S	STO+7	FIX 8
9	FRAC	STO+5	CHS	RCL 6	GTO 00

5. Operating Instructions

They are quite different for all three phases.

Phase I: Load program I, move to RUN. Load n into R_0. Press

PRGM
R/S

* When integer i (in FIX 0 notation) is shown, load b_i into the X-register and press

R/S

When the calculation stops, the number displayed is y_i and should be recorded. Press

R/S

to get back to * (i increasing from 0 to n). When 0.00 is shown, phase I is terminated.

Phase II: Load program II, move to RUN. Load n into R_0. Press

PRGM
R/S

* When integer i (in FIX 0 notation) is shown, load y_i (as computed in phase I) into the X-register and press

R/S

Unless $i = n$ (when $hs_n = y_n$ and therefore the calculator goes back to * right away and displays $i = n-1$) the number displayed at the next stop is $h \cdot s_i$ and should be recorded. Press

R/S

to get back to * (i decreasing from n to 0). When -1.00000000 is shown, phase II is terminated.

<u>Phase III:</u> Load program III, move to RUN. Load leftmost interpolating point x_0 into R_0 and h into R_1. Load the value of x, at which the spline interpolant g should be evaluated into the X-register. * Press

PRGM
R/S

At the first stop the integer i (corresponding to the interval, in which x lies) is shown in FIX 0 notation. Load

f_i into Y , hs_{i+1} into X

(as computed in phase II) and press

R/S

At the final stop g(x) is shown. If g has to be evaluated for another x, load new value of x into X and go back to *.

6. Example

Let $n := 20$, $x_k := -1 + h \cdot k$, $h = 0.1$, $k = 0,1,2,\ldots,20$, $f_{10} = 1$, $f_k = 0$ for $k \neq 10$, see Essentials, Demonstration 5.9-1. Input to

phase I: $n = 20$; $b_k = 0$ for $k \neq 9,11$, $b_9 = 3$, $b_{11} = -3$

II: $n = 20$ and results from I

III: $x_0 = -1$, $h = 0.1$ and results from II

Demonstrations 6.1-1, 6.3-1:

MIDPOINT, TRAPEZOIDAL, AND SIMPSON VALUES OF DEFINITE INTEGRAL

1. Purpose

Given a definite integral

$$I := \int_0^b f(x)dx , \qquad b > 0 ;$$

to compute, for n = 0, 1, 2, ..., its midpoint, trapezoidal and "Simpson" (i.e. first Romberg acceleration of trapezoidal value) values.

2. Method

According to Essentials, §6.1 (midpoint and trapezoidal) and §6.3 (including problem 8, Simpson). We recall the formulas to be implemented:

Midpoint: $$M_n := h_n \sum_{k=0}^{2^n-1} f((k+\tfrac{1}{2})h_n), \quad h_n := \frac{b}{2^n} ;$$

Trapezoidal: $$T_{n+1} := \frac{1}{2}(T_n + M_n) ;$$

Simpson: $$S_{n+1} := T^*_{n+1} := \frac{1}{3}(T_n + 2M_n) .$$

3. Flow Diagram

We put $M := M_n$, $T := T_{n+1}$, $T_0 := T_n$, $S := S_{n+1}$, $h := h_n$.

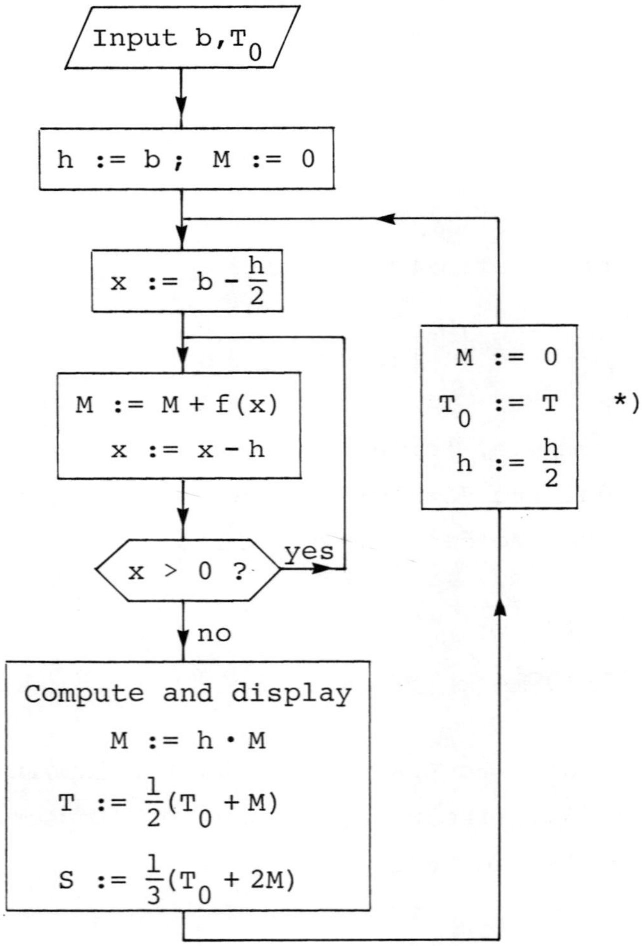

The sum in the formula for M, in terms of the sampling points used, is evaluated from right to left (rather than from left to right as indicated by $\sum$ from $k=0$ to $k=2^n-1$). In the program given in section 4 below the instructions in the box *), for reasons of economical use of the programming locations, are intertwined with the computation of M, T and S.

4. Storage and Program

R_0	R_1	R_2	R_3	R_4	R_5	R_6	R_7
b	b	x	T_0	0			
	h		T	M			

	00	10	20	30	40
0	╳	STO-2	+	STO-4	
1	RCL 0	RCL 2	2	+	
2	RCL 1	x > 0	STO÷1	3	
3	2	GTO 07	÷	÷	
4	÷	RCL 1	STO 3	GTO 00	
5	-	STO*4	R/S	→	
6	STO 2	RCL 3	R ↓		
7	→ GSB 35	RCL 4	RCL 4		
8	STO+4	R/S	+		
9	RCL 1	RCL 3	RCL 4		

Programming locations 35 ÷ 49 are available for the subroutine to compute f(x), assuming x in X and x in R_2. The subroutine must terminate with RTN. Storage registers $R_5 \div R_7$ may be used for additional storage.

5. Operating Instructions

Load program, including subroutine to compute f(x). Move to RUN and select mode of displaying numbers, e.g. by pressing

FIX 8

Load data as follows:

b = upper limit of integration
<u>both</u> into R_0 <u>and</u> R_1.

$T_0 := \frac{b}{2}(f(0) + f(b))$ into R_3.

0 into R_4.

Press

PRGM
R/S

to start the computation. *) After a while (longer as n increases) the calculator will display M_n. Upon pressing

R/S

T_{n+1} is computed (quickly) and displayed. After pressing

R/S

once more, S_{n+1} is computed (quickly) and displayed. To compute the next set of values M, T, and S, press

R/S

and again follow the instructions from *).

6. Example

$$f(x) := \frac{\sin x}{x}, \qquad b := \pi, \qquad f(0) := \lim_{x\to 0} f(x) = 1,$$

see Essentials, Demonstration 6.1-1 for values M and T and Demonstration 6.3-1 for values S. Subroutine to compute f(x):

```
35   RAD
36   sin
37   RCL 2
38   ÷
39   RTN
```

Demonstration 6.2-2:

STIRLING'S FORMULA

1. Purpose

To test the accuracy of Stirling's formula to approximate n! .

2. Method

Computation of $F := n!$, the Stirling approximation $S := \sqrt{2\pi n}\,(\frac{n}{e})^n$ and the quotient of the two, see Essentials, §6.2.

3. Flow Diagram

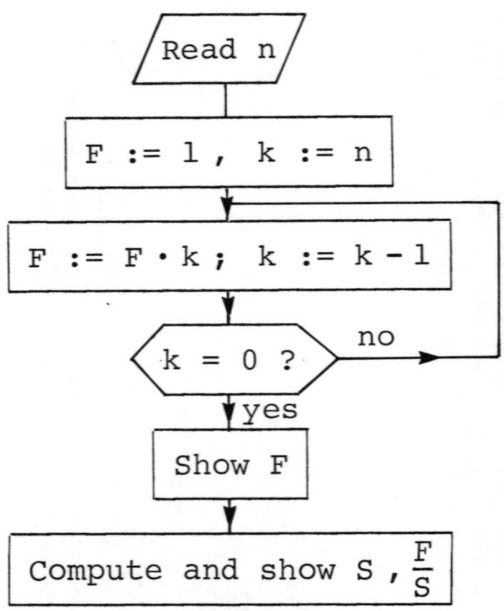

4. Storage and Program

R_0	R_1	R_2	R_3	R_4	R_5	R_6	R_7
n	k	F					

n∈X

	00	10	20	30	40
0	╳	STO−1	RCL 0	÷	
1	STO 0	GTO 05	π	GTO 00	
2	STO 1	→ RCL 2	*		
3	1	R/S	2		
4	STO 2	RCL 0	*		
5	→ RCL 1	1	$\sqrt{x}$		
6	x = 0	e^x	*		
7	GTO 12	÷	R/S		
8	STO*2	RCL 0	RCL 2		
9	1	y^x	$x \gtrless y$		

5. Operating Instructions

Load program, move to RUN and choose mode of displaying numbers, e.g. by pressing

SCI 6

Load n into X-register and press

PRGM
R/S

*) The computer will first compute and show n! , then, after pressing

R/S

the Stirling approximation and, after pressing

R/S

once more, their quotient. Load next value of n into X-register, press

R/S

and go to *), etc.

6. Examples

See Essentials, Demonstration 6.2-1. It can be verified numerically, that $\lim_{n\to\infty} \frac{F}{S} = 1$.

Problem 6.2-4:

PIPPENGER'S PRODUCT

R_0	R_1	R_2	R_3	R_4	R_5	R_6	R_7
$\sqrt{2}$	$\sqrt{2}$	p' old	p'		f	2	2m
p old	p					m	

	00	10	20	30	40
0		STO+6	GTO 07	STO 0	-
1	→ RCL 6	RCL 6	RCL 5	-	1
2	2	x^2	RCL 6	3	5
3	*	STO÷5	1/x	÷	÷
4	STO 7	1	y^x	-	-
5	1	STO+6	STO*1	STO 3	R/S
6	STO 5	RCL 6	RCL 1	R/S	GTO 01
7	→ RCL 6	STO*5	R/S	RCL 2	
8	STO*5	RCL 7	RCL 0	RCL 3	
9	1	x > y	RCL 1	STO 2	

Evaluates partial products of Pippenger's product,

$$p := \prod_{k=1}^{\infty} f_k ,$$

where

$$f_1 = \left(\frac{2}{1}\right)^{1/2}, \qquad f_k = \left(\frac{m}{m+1}\ \frac{m+2}{m+1}\ \frac{m+2}{m+3}\ \cdots\ \frac{2m}{2m-1}\right)^{\frac{1}{2m}},$$

$$m := 2^k,$$

and applies Romberg convergence acceleration twice.

Values are

p	p'	p"
1.		
1.373178096	1.359499607	
1.362672263	1.359170319	1.359148366
1.360025257	1.359142922	1.359141096
1.359362072	1.359141044	1.359140919
1.359196217	1.359140924	1.359140916
1.359154741	1.359140916	1.359140915

$$\frac{e}{2} = \qquad 1.359140914$$

Demonstrations 6.3-2 ÷ 6.3-4:

MIDPOINT RULE WITH STEP REFINEMENT AND TWO-FOLD ROMBERG CONVERGENCE ACCELERATION

1. Purpose

To approximate the integral

$$I := \int_0^b f(x)dx , \qquad b > 0 ,$$

by computing its midpoint values and accelerating their convergence.

2. Method

Midpoint rule (see Essentials, §6.1) and Romberg acceleration (see Essentials, §6.3). We compute the first three columns of the Romberg scheme by means of the formulas

$$M_n := h_n \prod_{k=0}^{2^n-1} f((k+\tfrac{1}{2})h_n) , \quad h_n := \frac{b}{2^n}, \quad n=0,1,2,...,$$

(midpoint values)

$$M_n^* := M_n - \frac{1}{3}(M_{n-1} - M_n) \ , \quad n=1,2,\ldots,$$

(first accelerated values)

$$M_n^{**} := M_n^* - \frac{1}{15}(M_{n-1}^* - M_n^*) \ , \quad n=2,3,\ldots,$$

(second accelerated values)

3. Flow Diagram

Though the first and second accelerated values are only defined for $n \geqq 1$ or $n \geqq 2$, respectively, the program given below does not treat the cases $n = 0$ and $n = 1$ separately. Instead it indiscriminately carries through all the formulas given in section 2 for $n = 0, 1, 2, \ldots$. This measure is not dangerous (e.g. no division by zero or accumulation of wrong values) and saves a lot of programming space, but it causes the second, third and sixth values after starting the program to be meaningless (see section 5). We put $h := h_n$, $M := M_n$, last $M := M_{n-1}$, $M^* := M_n^*$, last $M^* := M_{n-1}^*$, $M^{**} := M_n^{**}$.

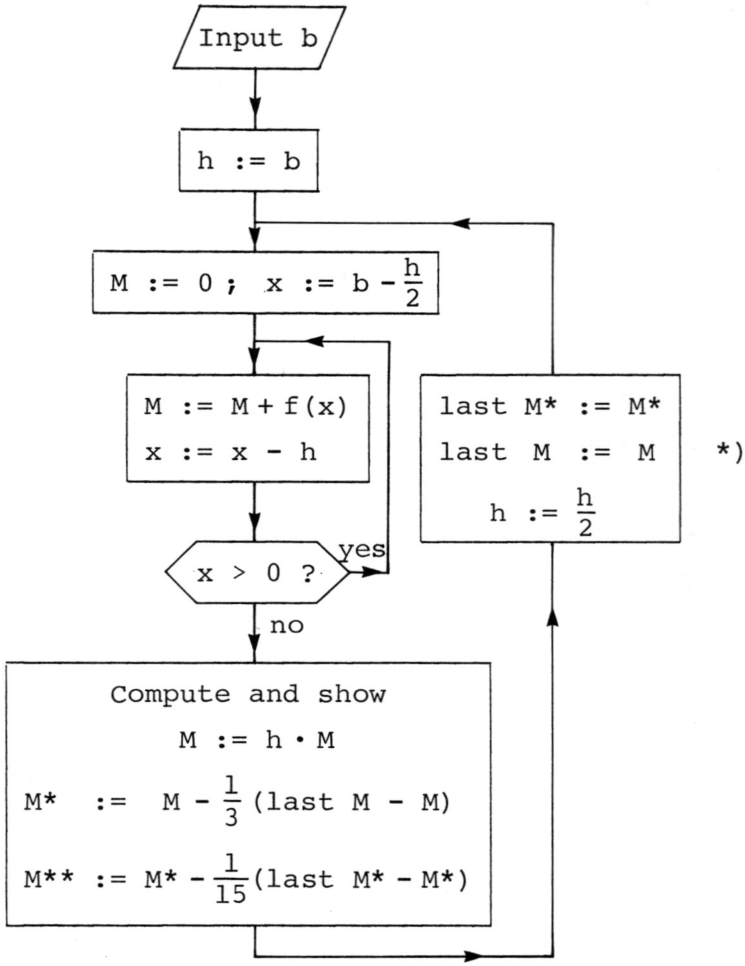

As in the preceding program (Midpoint, Trapezoidal, Simpson) the sum for M, with respect to the interval of integration, is evaluated from right to left. And the instructions in the box *) are intertwined with the computations of M, M* and M** to save programming space.

4. Storage and Program

R_0	R_1	R_2	R_3	R_4	R_5	R_6	R_7
b	b	x	M	last M	M*	last M*	

	00	10	20	30	40
0	X			STO÷1	STO 5
1	0		STO+3	RCL 3	R/S
2	STO 3		RCL 1	R/S	RCL 6
3	RCL 0		STO−2	RCL 4	RCL 5
4	RCL 1		RCL 2	RCL 3	STO 6
5	2		x > 0	STO 4	−
6	÷		GTO 09	−	1
7	−		RCL 1	3	5
8	STO 2		STO*3	÷	÷
9	→		2	−	−

Programming locations 09 ÷ 20 should be used for the program (not subroutine, to save space) to compute f(x). This program should assume x in X and x in R_2 and put f(x) into X. R_7 is available for additional storage.

5. Operating Instructions

Load program, including program to compute f(x). Move to RUN. Select mode of displaying numbers, e.g. by pressing

FIX 8

Load b, the upper limit of integration, both into R_0 and R_1. Press

PRGM
R/S

The calculator computes and displays the first midpoint value M_0. Upon pressing

R/S

a _meaningless_ number (M_0^* is _not_ defined) will be displayed. Pressing

R/S

once more produces another _meaningless_ number (M_0^{**} is _not_ defined). After pressing

R/S

again, the calculator will start the computation of the second row of the Romberg scheme and stop at the display of M_1. Press

R/S

to obtain M_1^*. After pressing

R/S

again, the last _meaningless_ number (M_1^{**} is _not_ defined) will be computed and shown. From now on all the numbers produced by the program are meaningful. Press

R/S

to reach the next row of the scheme. After a while (its length is proportional to 2^n) M_n is displayed. Press

R/S

to obtain M_n^* and

R/S

again, to obtain M_n^{**}. Proceed as indicated above to obtain the values of the next row. If only the second accelerated values are to be recorded, then instructions 32 and 41 can be changed to

PAUSE

in which case

R/S

has to be pressed only once, that is, after the computation of a whole row of the scheme.

6. Examples

[1] $f(x) := \frac{\sin x}{x}$, $b := \pi$, see Essentials, Demonstration 6.3-2. The program to compute f is

```
09   RAD
10   sin
11   RCL 2
12   ÷
13   NOP
 :    :
20   NOP
```

The fourth and fifth columns given in Essentials, have to be computed by hand using the formulas (now written without stars):

$$M_{n,3} := M_{n,2} - \frac{1}{63}(M_{n-1,2} - M_{n,2}), \quad n=3,4,\ldots ,$$

(third accelerated values)

$$M_{n,4} := M_{n,3} - \frac{1}{255}(M_{n-1,3} - M_{n,3}), \quad n=4,5,\ldots .$$

(fourth accelerated values)

[2] $f(x) := -\frac{\text{Log}(1-x)}{x}$, $b = 1$, see Essentials, Demonstration 6.3-3. Program to compute f:

```
09   CHS
10   1
11   +
```

```
12   LN
13   RCL 2
14   ÷
15   CHS
16   NOP
 :    :
20   NOP
```

[3] $f(x) := (\sqrt{1 - k^2\sin^2 x})^{-1}$, see Essentials, Demonstration 6.3-4. The program for f, assuming k^2 in R_7, is:

```
09   RAD
10   sin
11   x^2
12   RCL 7
13   *
14   CHS
15   1
16   +
17   √x
18   1/x
19   NOP
20   NOP
```

Before starting the program k^2 needs to be stored into R_7. The results for

$$b = \frac{\pi}{2} \quad \text{and} \quad k^2 = 0.5 \quad \text{or} \quad 0.99$$

and

$$b = 1 \quad \text{and} \quad k^2 = 0.5$$

can be looked up in Essentials.

Demonstration 6.3-5:

CONVERGENCE ACCELERATION IN THE COMPUTATION OF π AND OF OTHER CONSTANTS

1. Purpose

To demonstrate convergence acceleration in non-linear recurrence relations

$$y_{n+1} = y_n \sqrt{\frac{2}{1 + \frac{y_{n-1}}{y_n}}} ,$$

y_0 and y_1 given starting values.

2. Method

It can be shown that these recurrence relations satisfy an error law of the form

$$y_n = s + C_1 4^{-n} + C_2 4^{-2n} + \ldots + C_m 4^{-mn} + 0(4^{-(m+1)k}) ,$$

where s is the limit of the sequence $\{y_n\}$ and C_k, $k = 1, 2, \ldots, m$, are constants. Hence the Romberg algorithm (see Essentials, §6.3) may be applied. Along with y_n we generate four more columns of the Romberg scheme by:

$$y_n^* = y_n - \frac{1}{3}(y_{n-1} - y_n) ,$$

$$y_n^{**} = y_n^* - \frac{1}{15}(y_{n-1}^* - y_n^*) ,$$

$$y_n^{(3)} = y_n^{**} - \frac{1}{63}(y_{n-1}^{**} - y_n^{**}) ,$$

$$y_n^{(4)} = y_n^{(3)} - \frac{1}{255}(y_{n-1}^{(3)} - y_n^{(3)}) .$$

The values $y_n^{(m)}$ are meaningful only if $n \geqq m+1$.

3. Flow Diagram

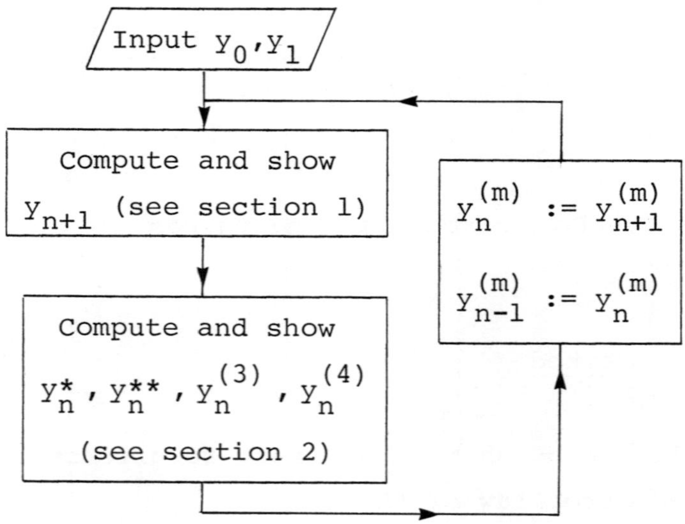

In order to save programming space the program given in section 4 below

a) for $n \geqq 2$ computes all the numbers $y_n^{(m)}$, i.e. generates a rectangular scheme and hence some meaningless numbers,

b) performs the shifts of the box on the right hand side of the flow diagram along with the computation of the $y_n^{(m)}$ (right after they have been used for the last time).

4. Storage and Program

R_0	R_1	R_2	R_3	R_4	R_5	R_6	R_7
y_0	y_1	y^*_{n-1}	y^*_n	y^{**}_{n-1}	y^*_n	$y^{(3)}_{n-1}$	$y^{(3)}_n$
y_{n-1}	y_n						

	00	10	20	30	40
0		STO*1	R/S (PAUSE)	R/S (PAUSE)	R/S (PAUSE)
1	2	RCL 1	RCL 2	RCL 4	RCL 6
2	RCL 0	R/S (PAUSE)	RCL 3	RCL 5	RCL 7
3	RCL 1	RCL 0	STO 2	STO 4	STO 6
4	STO 0	RCL 1	-	-	-
5	÷	-	1	6	2
6	1	3	5	3	5
7	+	÷	÷	÷	5
8	÷	-	-	-	÷
9	$\sqrt{x}$	STO 3	STO 5	STO 7	-

5. Operating Instructions

Load program, move to RUN, select mode of displaying numbers. Load starting values: y_0 into R_0, y_1 into R_1. Press

PRGM

R/S

to start the computation of the rows of the scheme. The numbers

$$y_2, \quad y_2^*, \quad y_2^{**}, \quad y_2^{(3)}, \quad y_2^{(4)}$$

$$y_3, \quad y_3^*, \quad y_3^{**}, \quad y_3^{(3)}, \quad y_3^{(4)}$$

$$\cdots$$

will be displayed row by row. After each display, press

R/S

to continue.

Note: In the display of the k-th row the numbers $y_k^{(m)}$ are meaningless if $m \geqq k$ because insufficient data are available for the construction of the complete Romberg scheme. Thus, only from the 5th row onward all numbers displayed are meaningful.

Since the sequence $\{y_4^{(m)}\}$, the last computed column of the Romberg scheme, converges to the limit most rapidly (see Essentials, Theorem 6.3), it is reasonable to change the instructions 12, 20, 30 and 40 to

PAUSE

Then y_n, y_n^*, y_n^{**}, $y_n^{(3)}$ will be shown briefly and the computer will stop at the display of $y_n^{(4)}$. Then

R/S

has to be pressed to start the computation of the next row.

6. Examples

[1] $y_0 = 0$, $y_1 = 1$, then $y_n = 2^n \sin(2^{-n}\pi)$ and $y_n \to \pi$, see Essentials, Demonstrations 1.3-4 (recurrence relation and limit) and 6.3-5 (numerical values).

[2] $y_0 = x\sqrt{1-x^2}$, $y_1 = x \quad (0 < x \leq 1)$, then $y_n = 2^{n-1}\sin(2^{1-n}\arcsin x)$ and $y_n \to \arcsin x$ (see Essentials, §1.3, problem 4*d).

If $x = 1$, then $y_0 = 0$, $y_1 = 1$, and hence all the columns of the scheme should converge to $\arcsin 1 = 1.57079633$. Indeed we obtain:

1.00000000				
1.41421356	1.55228475			
1.53073373	1.56957379	1.57072639		
1.56072258	1.57071886	1.57079520	1.57079629	
1.56827424	1.57079147	1.57079631	1.57079633	1.57079633

We see that y_5 has only one correct decimal digit, when $y_5^{(3)}$ and $y_5^{(4)}$ have converged already. Only y_{15} shows full accuracy.

If $x = \sqrt{2}/2$, then $y_0 = 1/2$ and $y_1 = \sqrt{2}/2$, and we have to exhibit convergence to $\arcsin \sqrt{2}/2 = \pi/4$ $= 0.78539816$. Results:

0.70710678			
0.76536686	0.78478689		
0.78036129	0.78535943	0.78539760	
0.78413712	0.78539573	0.78539815	0.78539816

Convergence has been reached already.

3 $y_0 = \frac{1}{2}(x - \frac{1}{x})$, $y_1 = \sqrt{x} - \frac{1}{\sqrt{x}}$ $(x > 0)$, then

$y_n = 2^{n-1}(x^{2^{-n}} - x^{-2^{-n}})$ and $y_n \to \operatorname{Log} x$

(see Essentials, §1.3, problem 4*c).

If $x = e$, then $y_0 = 1.17520119$ and $y_1 = 1.04219061$. The computation yields:

1.01044927 0.99986882
1.00260620 0.99999185 1.00000005
1.00065117 0.99999949 1.00000000 1.00000000 ,

indeed Log e = 1.

If $x = 10$, then $y_0 = 4.45$ and $y_1 = 2.84604989$, and we get:

2.43187617 2.29381826
2.33450889 2.30205313 2.30260212
2.31054929 2.30255209 2.30258536 2.30258509 = Log 10

Ten more iterations would be required for y_n to converge.

Problem 6.3-7:

PLANA INTEGRAL

R_0	R_1	R_2	R_3	R_4	R_5	R_6	R_7
h_0	$\frac{1}{2}h_0$	0					
$2^{-k}h_0$	y	s					

	00	10	20	30	40
0			1	RCL 0	RCL 2
1	→Program		-	STO+1	RCL 3
2	to		÷	GTO 01	RCL 2
3	place		RCL 2	→ RCL 0	STO-2
4	g(y)	RCL 1	x ≷ y	STO*2	STO 3
5	into X,	π	STO+2	STO 1	PAUSE
6	assuming	*	x ≷ y	2	-
7	$y \in R_1$	2	RCL 2	STO÷0	3
8		*	x = y	4	÷
9		e^x	GTO 33	STO÷1	-

Evaluates

$$I := \int_0^\infty \frac{g(y)}{e^{2\pi y} - 1}\, dy$$

using midpoint rule with steps h_0, $\frac{1}{2}h_0$, $\frac{1}{4}h_0$, ... and applying convergence acceleration once.

$$q(y) = \frac{2y}{1+y^2}$$

	s	s'
$h_0 = 1$	0.036207166	———
	0.066297263	0.076327295
	0.074582005	0.077343585

Demonstration 6.4-1:

CLASSICAL RUNGE-KUTTA METHOD:
INITIAL VALUE PROBLEMS FOR ORDINARY DIFFERENTIAL EQUATIONS

1. Purpose

To integrate numerically an ordinary differential equation (ODE)

$$y' = f(x,y)$$

subject to the initial condition

$$y(x_0) = y_0 \ .$$

2. Method

Classical Runge-Kutta method, see Essentials, §6.4. Choosing an integration step h, we determine approximate values y_n of the values $y(x_n)$ of the exact solution $y(x)$ at the points $x_n := x_0 + n \cdot h$, $n = 1, 2, \ldots$, by the formula

$$y_{n+1} = y_n + \frac{1}{3}(k_1 + 2k_2 + 2k_3 + k_4) \ ,$$

where

$$k_1 := \frac{h}{2} f(x_n, y_n)$$

$$k_2 := \frac{h}{2} f(x_n + \frac{h}{2} , y_n + k_1)$$

$$k_3 := \frac{h}{2} f(x_n + \frac{h}{2} , y_n + k_2)$$

$$k_4 := \frac{h}{2} f(x_n + h , y_n + 2k_3) .$$

3. Flow Diagram

The very ingenious program given in section 4 below is due to Dr. Oliver Pretzel.
It basically follows the pattern of the flow diagram for the classical Runge-Kutta method given in Essentials, Fig. 6.4d. However, in order to use one integer, once it is available in the X-register, for more than one purpose, the author lets the indices of the k_j range from -1 to 2 rather than 1 to 4. Together with putting $j := j+1$ before the last branching point, this enables him to compute k_1 and k_2 with $y^* = y + j \cdot k_j$. Furthermore the denominator 3 in the formula for y_{n+1} can also be used to compute $j := j - 3$, i.e. to re-set $j := -1$ before starting the algorithm for a new x.

4. Storage and Program

R_0	R_1	R_2	R_3	R_4	R_5	R_6	R_7
$\frac{h}{2}$	-1	0	x_0	y_0			
	j	s	x	y			

$y_0 \in X$

	00	10	20	30	40
0	⊠		1	STO 1	GTO 01
1	→		RCL 1	x = 0	→ RCL 3
2			x > y	GTO 37	PAUSE
3			GTO 41	x ⋛ y	RCL 2
4			x = 0	STO+2	STO-2
5			GTO 29	*	3
6			RCL 0	GTO 38	STO-1
7		RCL 0	STO+3	→ R ↓	÷
8		*	R ↓	→ RCL 4	STO+4
9		STO+2	→ +	+	RCL 4

Instructions 01 ÷ 16 should be used for the program to compute f(x,y*), assuming y* in X and x in R_3; this program should put f(x,y*) into the X-register. Registers R_5, R_6, R_7 are available for temporary or permanent storage. In this form (i.e. without using subroutines) the program can even be used on the HP-25 (after interchanging instructions 20 and 21 and changing instruction 22 to x < y).

5. Operating Instructions

Load program, including program to compute f(x,y*). Move to RUN. Select mode of displaying numbers and load data as follows:

$\frac{h}{2}$	into	R_0
-1	into	R_1
0	into	R_2
x_0	into	R_3
y_0	into	R_4 ,

and leave y_0 in the X-register (if data are input in indicated order, it is already there). Press

PRGM
R/S

to start the computation. After a while x_1 will be shown briefly and the calculator will stop at the display of y_1. Press

R/S

to obtain x_2 and y_2, etc.

6. Example

$y' = x^2 - y^2$, $x_0 = 0$, $y_0 = 1$, see Essentials, Demonstration 6.4-1. Here the function to compute f(x,y*) is:

```
01    x^2
02    CHS
03    RCL 3
04    x^2
05    +
06    NOP
 :     :
16    NOP
```

The accelerated values (by a variant of the Runge-Kutta algorithm) y*(h) and y**(h) can be computed "by hand".

Demonstration 6.5-1:

SIMPLIFIED RUNGE-KUTTA METHOD FOR SYSTEMS OF 2 EQUATIONS

1. Purpose

To integrate numerically a system of two first order differential equations

$$y' = f(x,y,z)$$
$$z' = g(x,y,z) ,$$

subject to the initial conditions

$$y(x_0) = y_0$$
$$z(x_0) = z_0 .$$

Since a single second order differential equation

$$y'' = g(x,y,y') ,$$

subject to the initial conditions

$$y(x_0) = y_0$$
$$y'(x_0) = z_0 ,$$

by introducing $z := y'$ can be reformulated as system of two first order ODEs, it can be integrated with the same algorithm.

2. Method

Simplified Runge-Kutta method for systems of first order ODEs, see Essentials, §6.5. We recall the formulas for a system of 2 equations:

$$y_{n+1} = [y_n + \frac{h}{2}f(x_n,y_n,z_n)]$$

$$+ \frac{h}{2}f(x_n+h,y_n+hf(x_n,y_n,z_n),z_n+hg(x_n,y_n,z_n))$$

$$z_{n+1} = [z_n + \frac{h}{2}g(x_n,y_n,z_n)]$$

$$+ \frac{h}{2}g(x_n+h,y_n+hf(x_n,y_n,z_n),z_n+hg(x_n,y_n,z_n))$$

3. Flow Diagram

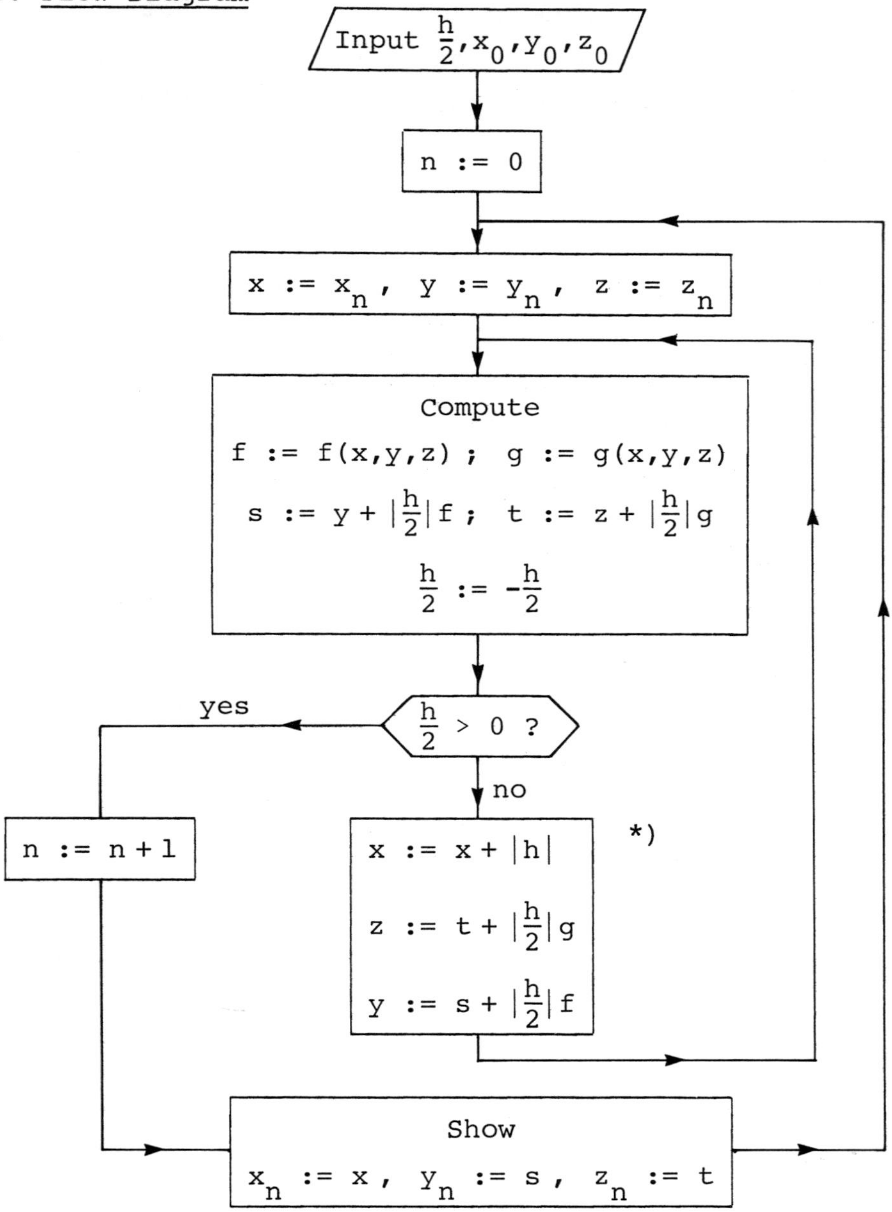

This flow diagram is based on the assumption that $h > 0$ (forward integration). During one iteration the branching point "$\frac{h}{2} > 0$?" is reached twice. First the answer will be "no", and the calculator continues by computing the arguments of f and g on the right hand side of the formula (see section 2); the absolute value bars in the box *) are for clarity only, since x - expressed in terms of the input h - is computed as $x - (-\frac{h}{2}) - (-\frac{h}{2})$, i.e. with the "new" $\frac{h}{2}$, whereas z and y (by a R↓ instruction) are computed with the "old" $\frac{h}{2}$. When the second time the answer is "yes", the appropriate sums to obtain y_{n+1} and z_{n+1} have been computed. The program does not actually count the iterations (and hence not carry out "n := 0" or "n := n+1", as indicated in the flow diagram).

4. Storage and Program

R_0	R_1	R_2	R_3	R_4	R_5	R_6	R_7
$\frac{h}{2}$	x_0	y_0	z_0				
$-\frac{h}{2}$		s	t				
$+\frac{h}{2}$	x_n	y_n	z_n				

$y_0 \in Y$, $z_0 \in X$

	00	10	20	30	40
0	X			RCL 0	x ≷ y
1	→			CHS	RCL 2
2			RCL 0	STO 0	+
3			ABS	x > 0	x ≷ y
4			*	GTO 45	GTO 01
5			STO+2	STO-1	→ RCL 1
6			x ≷ y	STO-1	PAUSE
7			LAST X	R ↓	RCL 2
8			*	RCL 3	R/S
9			STO+3	+	RCL 3

Instructions 01 ÷ 21 should be used for the program computing f(x,y,z) and g(x,y,z):

Assume z in X , y in Y
Put f into X , g into Y

Registers $R_4 \div R_7$ are available for temporary or permanent storage.

5. Operating Instructions

Load program, including program computing f and g. Move to RUN, select mode of displaying numbers. Load data as follows:

$\frac{h}{2}$	into	R_0	(h must be > 0)
x_0	into	R_1	
y_0	into	R_2	
z_0	into	R_3	

and put y_0 into the Y-register and z_0 into the X-register of the stack (if data are input in indicated order, they are already there); this must not be forgotten! Press

PRGM
R/S

to start the computation. The calculator will briefly display x_1 and stop at the display of y_1. Press

R/S

to obtain z_1. * Upon pressing

R/S

the next iteration is started; x_n is displayed briefly, y_n indefinitely, and after pressing

R/S

again, z_n is displayed. Continue with *, etc.

6. Example

Oscillations of a mathematical pendulum of length L and mass m, see Essentials, Demonstration 6.5-1. Resulting first order system:

$$y' = z \qquad =: f(x,y,z)$$

$$z' = -\frac{g}{L}\sin y =: g(x,y,z)$$

with initial conditions

$$y(0) = y_0$$

$$y'(0) = 0 \qquad \text{(pendulum is released, not pushed)} \;,$$

where the gravitational constant $g = 9.81$ (m/sec^2). We set $L := 1$ and store the constant g into R_4. Hence the program computing the functions f and g can be written

```
01   x ≷ y
02   sin
03   RCL 4
04   CHS
05   *
06   x ≷ y
07   GTO 22
09   NOP
 :    :
21   NOP
```

Numerical example: $y_0 := \pi/3$. Therefore

RAD

needs to be pressed before starting the computation. We initially set

$$h = 0.0125$$

and obtain

x_n	y_n	z_n
0.000000	1.047198	0.000000
0.012500	1.046534	-0.106196
0.025000	1.044543	-0.212311
....	....	....
0.100000	1.004885	-0.842598
....	....	....
0.300000	0.680345	-2.332724
....	....	....
0.500000	0.119155	-3.109734
0.512500	0.080192	-3.121939
0.525000	0.041106	-3.129374
<u>0.537500</u>	<u>0.001957</u>	<u>-3.132015</u>
0.550000	-0.037194	-3.129856

Now $y_n < 0$ for the first time. Since our program only works with $h > 0$ we set $h := \frac{h}{2} = 0.00625$ and use the underlined data as input (remember, that y_n in Y and z_n in X). This yields

0.543750 -0.017618 -3.131535 ,

again a negative y_n. We repeat this process. Now $h = 0.003125$, yielding

0.540625	-0.007831	-3.131925 ,

i.e. a $y_n < 0$. Also $h = 0.001563$ and $h = 0.000781$ are unsuccessful. But $h = 0.000391$ yields

0.537891	0.000734	-3.132020
0.538281	-0.000490	-3.132021

With $h := \frac{h}{2} = 0.000195$ and the underlined data as input there results

0.538086	0.000122	-3.132021
0.538282	-0.000489	-3.132020

With $h = 0.000098$ and $h = 0.000049$ and the underlined data we still obtain negative values y_n, but $h = 0.000024$ leads to

0.538110	0.000046	-3.132021
0.538135	-0.000031	-3.132021

Putting $h := \frac{h}{4} = 0.000006$ and using the underlined data as input yields

0.531160	0.000027
0.538122	0.000008
0.538128	-0.000011

Always continuing in the same manner, we finally get

0.538125 0.000000

and hence the period T of the pendulum

$$T = 4x_n = 2.1525 .$$

Problem 6.5-1:

PREY-PREDATOR MODEL

R_0	R_1	R_2	R_3	R_4	R_5	R_6	R_7
$\frac{h}{2}$	x_0	y_0	z_0	a	b	c	d

	00	10	20	30	40
0		*	NOP		
1	$x \gtrless y$	R ↓	NOP		
2	ENTER	*			
3	ENTER	RCL 5	continue		
4	RCL 6	*	as in		
5	*	CHS	Demo.		
6	RCL 7	$x \gtrless y$	6.5-1		
7	-	RCL 4			
8	$x \gtrless y$	*			
9	R ↓	+			

Solves system

$$y' = ay - byz = y(a - bz)$$
$$z' = cyz - dz = z(dy - d)$$

by simplified Runge-Kutta method.

Demonstration 6.6-1:

DIFFERENTIAL EQUATION OF FIRST ORDER: TRAPEZOIDAL METHOD

1. Purpose

To integrate numerically an ordinary differential equation (ODE)

$$y' = f(x,y)$$

subject to the initial condition

$$y(x_0) = y_0 \ .$$

2. Method

Trapezoidal method, see Essentials, §6.6, i.e. choose integration step h and determine approximate values y_n to $y(x_n)$ of the exact solution $y(x)$ at the points $x_n = x_0 + n \cdot h$. The y_n, $n = 1, 2, \ldots,$ are computed by iteration according to

$$y_{n+1}^{(0)} = y_n + \frac{h}{2} f(x_n, y_n)$$

$$y_{n+1}^{(m+1)} = y_n + \frac{h}{2}\{f(x_n, y_n) + f(x_{n+1}, y_{n+1}^{(m)})\} \ ;$$

m = 0, 1, ...; the iteration is stopped when two successive iterates differ by less than ε. Then

$$y_{n+1} := y_{n+1}^{(m+1)} .$$

3. Flow Diagram

See Essentials, Fig. 6.6a.

4. Storage and Program

R_0	R_1	R_2	R_3	R_4	R_5	R_6	R_7
h	ε	x_0	y_0	$\frac{h}{2}f_0$			
		x_n	y_n	k			

	00	10	20	30	40
0	☒	*	RCL 2		
1	RCL 4	RCL 4	PAUSE		
2	STO-4	$x \gtrless y$	RCL 3		
3	STO+3	STO 4	GTO 00		
4	RCL 0	-	→		
5	STO+2	STO-3			
6	→ GSB 24	ABS			
7	RCL 0	RCL 1			
8	2	$x \leq y$			
9	÷	GTO 06			

Instructions 24 ÷ 49 should be used for the subroutine computing f(x,y). This subroutine should assume x in R_2 and y in R_3 and leave f(x,y) in the X-register. Its last instruction should be RTN.

5. Operating Instructions

Load program, including subroutine for f. Move operating switch to RUN. Choose mode of displaying numbers, e.g. by pressing

FIX 6

Load data as follows:

Integration step	h	into	R_0
Tolerance	ε	into	R_1
Starting values	x_0	into	R_2
	y_0	into	R_3
	$\frac{h}{2}f_0 := \frac{h}{2}f(x_0,y_0)$	into	R_4

Press

PRGM
R/S

to start the computation. After a while x_1 will be displayed briefly, and the calculator will stop at the display of y_1. Press

R/S

to obtain x_2 and y_2, etc.

6. <u>Example</u>

$y' = x^2 - y^2 =: f(x,y)$, $y(0) = 1$, see Essentials, Demonstration 6.6. Subroutine to compute f:

```
24   RCL 2
25   x²
26   RCL 3
27   x²
28   -
29   RTN
```

Input of $\varepsilon := 10^{-9}$:

```
EEX
9
CHS
STO 1
```

$x_0 = 0$, $y_0 = 1$, $f_0 = -1$. With integration steps h = 0.2, 0.1, 0.05 (step is successively halved) approximations T_1, T_2, T_3 to y(2) are computed. In view of the application of Romberg acceleration, it is reasonable to press

```
FIX 9
```

before computing the final value for each h. There results

$T_1 = 1.677871265$
$T_2 = 1.679066690$
$T_3 = 1.679361196$

With the formulas

$$T_n^* = T_n - \frac{1}{3}(T_{n-1} - T_n) , \qquad n = 2, 3,$$

$$T_3^{**} = T_n^* - \frac{1}{15}(T_2^* - T_3^*) ,$$

the accelerated values can be found.

Problem 6.6-5:

SIMPSON METHOD FOR $y' = f(x,y)$

R_0	R_1	R_2	R_3	R_4	R_5	R_6	R_7
h	ε	x_1	y_0	y_1	$\frac{h}{3}f_0$	$\frac{h}{3}f_1$	

	00	10	20	30	40
0		STO 5	$f(x_1,y_1)$	putting	STO 6
1	RCL 5	4	assuming	f	-
2	RCL 3	*	x_1	into	STO-4
3	+	STO+4	in	X	ABS
4	RCL 4	RCL 0	R_2	RCL 0	RCL 1
5	STO 3	STO+2	and	3	$x \leq y$
6	-	→ Program	y_1	÷	GTO 16
7	STO+4	to	in	*	RCL 2
8	RCL 6	com-	R_4	RCL 6	PAUSE
9	STO-6	pute	and	$x \gtrless y$	RCL 4

To start press

PRGM

R/S

Problem 6.6-7:

NUMEROV METHOD FOR $y'' = f(x,y)$

R_0	R_1	R_2	R_3	R_4	R_5	R_6	R_7
h	ε	x_1	y_0	y_1	$\frac{h^2}{12}f_0$	$\frac{h^2}{12}f_1$	

	00	10	20	30	40
0		STO 5	com-	into	STO 6
1	RCL 5	1	pute	X	-
2	RCL 3	0	f(x,y)	RCL 0	STO-4
3	-	*	assuming	x^2	ABS
4	RCL 4	STO+4	$x \in R_2$	1	RCL 1
5	STO 3	RCL 0	and	2	$x \leqq y$
6	+	STO+2	$y \in R_4$	÷	GTO 17
7	STO+4	→ Pro-	and	*	RCL 2
8	RCL 6	gram	putting	RCL 6	PAUSE
9	STO-6	to	f	$x \gtrless y$	RCL 4

To start press

PRGM

R/S

Demonstration 7.1-1:

SLOW FOURIER TRANSFORM, REAL DATA, $n = 8$, PERIOD 1

1. Purpose

Given $n = 2m = 8$ real data f_j stemming from a periodic function f with period $T = 1$. To compute the coefficients c_k, $k = -4, -3, \ldots, 4$, of the balanced trigonometric polynomial of degree $m = 4$, interpolating f (see Essentials, §7.1 for definitions, Theorem 7.1a for Existence and Uniqueness).

2. Method

Computation of the "ordinary", i.e. slow Fourier Transform, see Essentials, §7.1. Since for real data f_j, $c_{-k} = \overline{c_k}$, $k = 0, 1, \ldots, 4$, only $c_0, c_1, \ldots, c_4$ need to be computed by

$$c_k = \frac{1}{8} \sum_{j=0}^{7} f_j \cdot z_1^{-kj} , \qquad \text{with} \quad z_1 := e^{-i\frac{\pi}{4}} .$$

For the implementation on the calculator this can be written in a Horner-like form

$$c_k = \frac{1}{8}\left\{f_0 + z(f_1 + z(f_2 + \ldots z(f_6 + zf_7)\ldots))\right\} ,$$

where $z := z_1^{-k}$. Define

$$c_k =: c'_k + i \cdot c''_k \ .$$

It can be shown that c_0 and c_4 are real, i.e. $c''_0 = c''_4 = 0$.

3. Flow Diagram

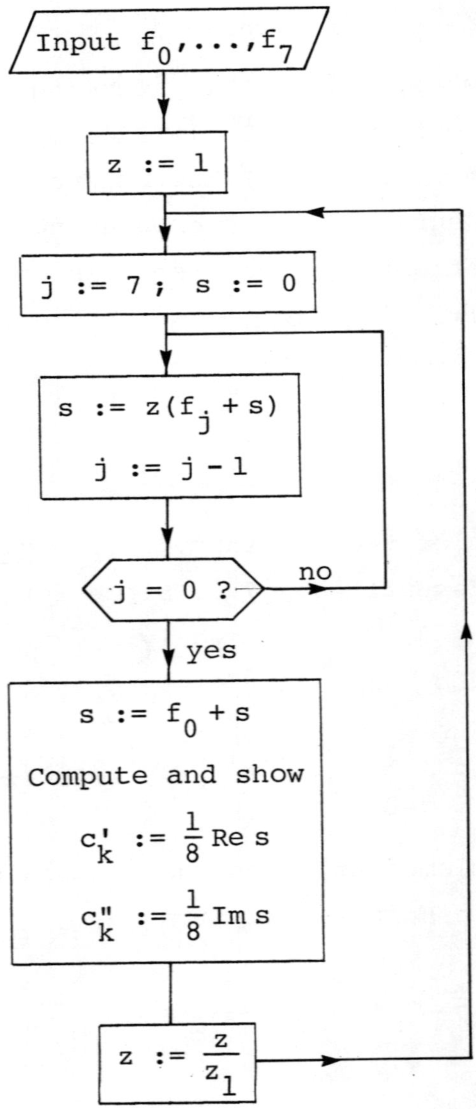

The index j has been used in the flow diagram for clarity only; it is not actually used in the program, given in section 4 below. Furthermore, since the z, always being roots of unity, all have modulus 1, the program only works with their arguments. The multiplication $z(f_j + s)$ is implemented in the subroutine starting with instruction 41. This subroutine first transforms $f_j + s$ into polar coordinates, then multiplies the two factors by adding the arguments (and does not multiply $1 \cdot |f_j + s|$) and transforms back to cartesian coordinates. Since all the storage registers are occupied by the f_j, the stack has to be used extensively.

4. Storage and Program

R_0	R_1	R_2	R_3	R_4	R_5	R_6	R_7
f_0	f_1	f_2	f_3	f_4	f_5	f_6	f_7

	00	10	20	30	40
0	╳	GSB 41	RCL 1	÷	GTO 03
1	RAD	RCL 4	+	R/S	→ P
2	STK	+	GSB 41	R ↓	R ↓
3	RCL 7	GSB 41	RCL 0	π	+
4	GSB 41	RCL 3	+	4	R ↓
5	RCL 6	+	8	÷	R ↓
6	+	GSB 41	÷	-	CLX
7	GSB 41	RCL 2	R/S	ENTER	+
8	RCL 5	+	R ↓	ENTER	→ R
9	+	GSB 41	8	0	RTN

5. Operating Instructions

Load program, move to RUN. Select mode of displaying numbers, e.g. by pressing

FIX 8

Press

PRGM
R/S

to start the computation. If

R/S

is pressed after each display, the results are obtained in the following order:

c'_0 , c''_0

c'_1 , c''_1

c'_2 , c''_2

c'_3 , c''_3

c'_4 , c''_4

6. Example

Daily temperatures in Johnson City, New Mexico, see Essentials, Demonstration 7.1-1.

Demonstration 7.1-1:

FOURIER SYNTHESIS, REAL DATA, $n = 8$, PERIOD 1

1. Purpose

Given the coefficients $c_k := c_k' + ic_k''$, $k = 0,1,\ldots,4(=m)$, of the balanced trigonometric polynomial $t(\tau)$ of degree 4, interpolating a real valued function f, given at $n = 8 = 2m$ points τ_k, $k = 0,1,\ldots,n-1$. To evaluate $t(\tau)$ for arbitrary τ, see Essentials, §7.1 and previous program "Slow Fourier Transform".

2. Method

Evaluation of

$$t(\tau) := \sum_{k=-3}^{3} c_k e^{2\pi ik\tau} + \frac{1}{2}(c_4 e^{2\pi i4\tau} + c_{-4}e^{-2\pi i4\tau})$$

$$= c_0 + 2\,\mathrm{Re}\{\sum_{k=1}^{3} c_k e^{2\pi ik\tau} + \frac{1}{2}c_4 e^{2\pi i4\tau}\},$$

$c_k := c_k' + ic_k''$, by Horner's rule, see Essentials, §7.1. The fact, that $c_0'' = c_4'' = 0$, in the case of real data, is taken into account before starting the computation.

3. Flow Diagram

Straightforward implementation of Horner's algorithm. Similar remarks as in the previous program "Slow Fourier Transform" hold with respect to the arithmetic of complex numbers and the use of the stack.

4. Storage and Program

R_0	R_1	R_2	R_3	R_4	R_5	R_6	R_7
c_0'	c_1'	c_1''	c_2'	c_2''	c_3'	c_3''	c_4'

$\tau \in X$

	00	10	20	30	40
0		2	RCL 3	RCL 2	$x \gtrless y$
1	RAD	÷	+	+	R ↓
2	2	→ R	$x \gtrless y$	$x \gtrless y$	$x \gtrless y$
3	*	RCL 5	RCL 4	GSB 39	R ↓
4	π	+	+	2	R ↓
5	*	$x \gtrless y$	$x \gtrless y$	*	R ↓
6	ENTER	RCL 6	GSB 39	RCL 0	+
7	ENTER	+	RCL 1	+	$x \gtrless y$
8	ENTER	$x \gtrless y$	+	GTO 00	→ R
9	RCL 7	GSB 39	$x \gtrless y$	→ P	RTN

5. Operating Instructions

Load program, move to RUN, choose mode of displaying numbers. Load data as follows:

c'_0 into R_0

c'_1 into R_1

c''_1 into R_2

c'_2 into R_3

c''_2 into R_4

c'_3 into R_5

c''_3 into R_6

c'_4 into R_7

(and do <u>not</u> load c''_0 and c''_4, which are zero). Load the value τ, at which t should be evaluated, into the X-register. Press

PRGM
R/S

The calculator will stop at the display of $t(\tau)$. If the polynomial has to be evaluated for another τ, load new value of τ into X and press

R/S

etc.

6. Example

Daily temperatures in Johnson City, New Mexico, see Essentials, Demonstration 7.1-1. Input to this program are the coefficients c_k, computed by means of the previous program "Slow Fourier Transform". Now we can compute the temperature at any other time h of the day, by letting τ = fraction of the day, elapsed at time h. (This has to be done, since the period T is assumed to be 1, i.e. here 1 day.)

Demonstration 7.2-1:

UNSTABLE TRIGONOMETRIC INTERPOLATION, n EVEN

1. Purpose

To demonstrate that the representation

$$t(\tau) = \frac{\sin n\pi\tau}{n} \sum_{k=0}^{n-1} (-1)^k \cot[\pi(\tau-\tau_k)] \cdot f_k$$

for the balanced trigonometric polynomial interpolating a real-valued function f with period 1, given at the points

$$\tau_k := \frac{k}{n}, \qquad k = 0, 1, \ldots, n-1,$$

$f_k := f(\tau_k)$, n even, is unstable, see Essentials, §7.2.

2. Method

Using the facts, that $\cot\alpha = \frac{1}{\tan\alpha}$, $\tan(\alpha-\pi) = \tan\alpha$, $f(0) = f(1)$ and letting $\tau_n := 1$, we can write

$$t(\tau) = \frac{\sin n\pi\tau}{n} \sum_{k=1}^{n} (-1)^k \frac{f(\tau_k)}{\tan[\pi(\tau-\tau_k)]} .$$

3. Flow Diagram

Straightforward. The summation is started with the term for $k = n$. The program given in section 4 below assumes, that f is given by a formula (rather than data points only).

4. Storage and Program

R_0	R_1	R_2	R_3	R_4	R_5	R_6	R_7
[n]	k	τ	ε_k	s	τ_k		

$\tau \in x$

	00	10	20	30	40
0	⊠	x = 0	*	GTO 09	GTO 00
1	RAD	GTO 31	tan	→ RCL 2	→
2	STO 2	RCL 0	÷	π	
3	RCL 0	÷	RCL 3	*	
4	STO 1	STO 5	*	RCL 0	
5	0	GSB 41	STO+4	STO÷4	
6	STO 4	RCL 2	1	*	
7	1	RCL 5	STO-1	sin	
8	STO 3	-	CHS	STO*4	
9	→ RCL 1	π	STO*3	RCL 4	

Instructions 41 ÷ 49 are available for the subroutine computing $f(\tau_k)$. This subroutine should assume τ_k in X and τ_k in R_5, leave f_k in X and terminate with the instruction RTN. Storage registers R_6 and R_7 can be used for additional storage.

5. Operating Instructions

Load program, including subroutine computing f_k. Move to RUN, select mode of displaying numbers. Load n (even) into R_0. Load τ, at which t should be evaluated, into X. Press

PRGM
R/S

The calculator will stop at the display of $t(\tau)$. If the polynomial has to be evaluated for another τ, load new value of τ into X and press

R/S

etc.

6. Example and Timing

We try the behavior of the representation of $t(\tau)$, given in section 1, for $f(\tau) \equiv 1$, see Essentials, Demonstration 7.2-1. Subroutine for f:

```
41   1
42   RTN
```

Time used to compute one value $t(\tau)$ approx. 27 sec.

Demonstration 7.2-2:

TRIGONOMETRIC INTERPOLATION, BARYCENTRIC FORMULA, PERIOD 1

1. Purpose

To evaluate the balanced trigonometric polynomial $t(\tau)$ of degree m interpolating a real-valued function f with period 1, given at the points

$$\tau_k := \frac{k}{n}, \qquad k = 0, 1, \ldots, n-1,$$

by means of the barycentric formula, which is a stable representation for $t(\tau)$, see Essentials, §7.2.

2. Method

If $n = 2m$ is even, we use the facts that $\cot\alpha = \frac{1}{\tan\alpha}$, $\tan(\alpha-\pi) = \tan\alpha$, $f(0) = f(1)$ and let $\tau_n := 1$, we obtain

$$t(\tau) = \frac{\sum_{k=1}^{n} (-1)^k \frac{f(\tau_k)}{\tan[\pi(\tau-\tau_k)]}}{\sum_{k=1}^{n} (-1)^k \frac{1}{\tan[\pi(\tau-\tau_k)]}},$$

see Essentials, Theorem 7.2a.

If $n = 2m + 1$ is odd, for similar reasons the foregoing formula with "tan" replaced by "sin" results.

3. Flow Diagram

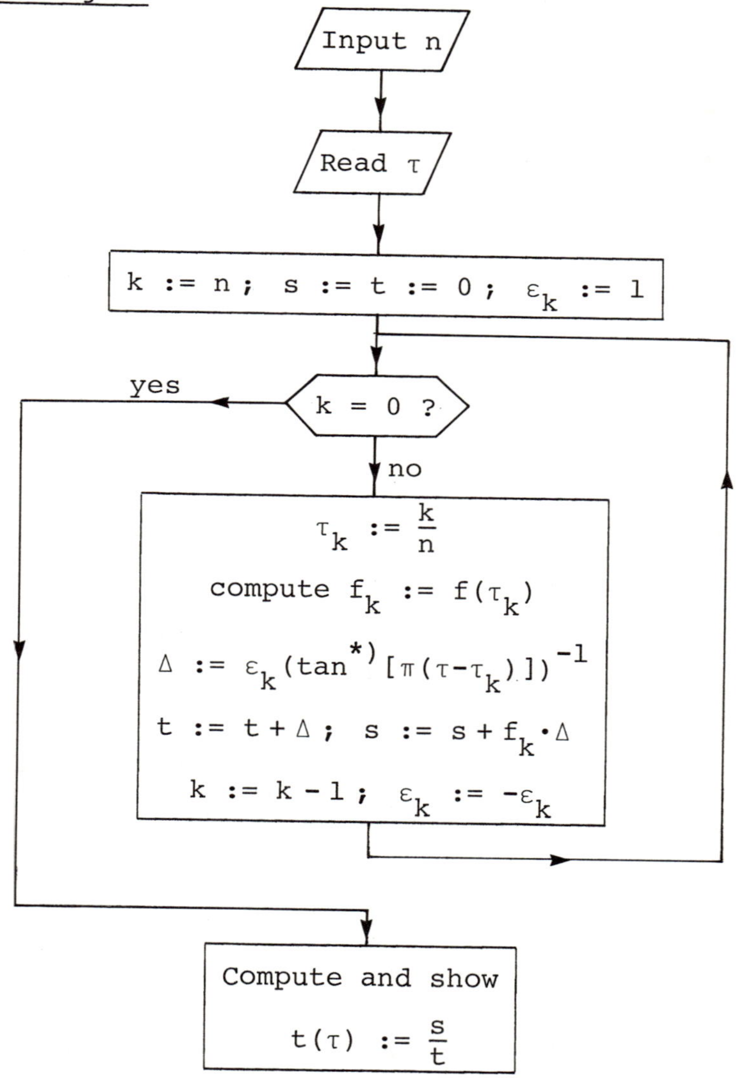

*) If n is odd, "tan" has to be replaced by "sin". The program given in section 4 below assumes, that f is given by a formula.

4. Storage and Program

R_0	R_1	R_2	R_3	R_4	R_5	R_6	R_7
n (boxed)	k	τ	τ_k	ε_k	s	t	

$\tau \in X$

	00	10	20	30	40
0	╳	→ RCL 1		RCL 2	*
1	RAD	x = 0		RCL 3	STO+5
2	STO 2	GTO 47		-	1
3	RCL 0	RCL 0		π	STO-1
4	STO 1	÷		*	CHS
5	0	STO 3		tan (sin)	STO*4
6	STO 5			1/x	GTO 10
7	STO 6			RCL 4	→ RCL 5
8	1			*	RCL 6
9	STO 4			STO+6	÷

Instructions 16 ÷ 29 should be used for the <u>program</u> to evaluate $f_k := f(\tau_k)$, assuming τ_k in X and in R_3 and putting f_k into X. R_7 is available for additional storage. Instruction 35 is "tan" if n is even and "sin" if n is odd.

5. Operating Instructions

Load program, including program to compute f_k. Move to RUN. Load n into R_0 and τ into the X-register. Press

```
FIX 9
PRGM
R/S
```

to obtain $t(\tau)$. If t needs to be evaluated for another τ, then load new value of τ into X and press

```
R/S
```

etc.

6. Example and Timing

Let $f(\tau) := \cos n\pi\tau$, $n = 12$, see Essentials, Demonstration 7.2-2. Program computing f_k:

```
16    RCL 0
17    *
18    π
19    *
20    cos
21    NOP
⋮      ⋮
29    NOP
```

Time needed to compute one value $t(\tau)$ approx. 37 sec.

Demonstrations 7.2-3, 7.4-1:

DYNAMIC TRIGONOMETRIC INTERPOLATION, PERIOD 1

1. Purpose

Given a real-valued periodic function f with period 1. To compute, for $\ell = 0, 1, 2, \ldots,$ the balanced trigonometric polynomials $t_\ell(\tau)$ interpolating f at the $n := 2^\ell$ points $\tau_k := k/n,\ k = 0, 1, 2, \ldots, n-1$.

2. Method

The $t_\ell(\tau)$ are generated recursively by

$$t_\ell(\tau) := \frac{a_\ell - r_\ell}{b_\ell - s_\ell},$$

where

$$a_0 = f(0) \cot \pi\tau, \quad b_0 = \cot \pi\tau, \quad r_0 = s_0 = 0$$

and

$$a_{\ell+1} = a_\ell + r_\ell, \quad b_{\ell+1} = b_\ell + s_\ell,$$

$$r_\ell = \sum_{\substack{k=1 \\ k \text{ odd}}}^{n-1} f(\tau_k) \cot \pi(\tau - \tau_k),$$

$$s_\ell = \sum_{k=1}^{n-1} \cot \pi(\tau-\tau_k) ,$$

see Essentials, Theorem 7.2b.

3. Flow Diagram

The program given in section 4 below basically follows the pattern of the flow diagram in Fig. 7.2a of Essentials. Thus the sums in the formulae for r_ℓ and s_ℓ are evaluated from $k = n-1$ downward.

4. Storage and Program

R_0	R_1	R_2	R_3	R_4	R_5	R_6	R_7
τ		1	1	0	0	0	0
$\tau \in X$							

	00	10	20	30	40
0	╳		-	RCL 3	STO-7
1	RAD		π	x > 0	STO+5
2	STO 0		*	GTO 05	-
3	→ 1		tan	RCL 4	÷
4	STO-3		1/x	RCL 6	R/S
5	→ $\bar{x}$		STO+7	STO-6	RCL 2
6		RCL 0	*	STO+4	STO+2
7		$\bar{x}$	STO+6	-	RCL 2
8		x ⪌ y	2	RCL 5	STO 3
9		R ↓	STO-3	RCL 7	GTO 03

Instructions 06 ÷ 15 should be used for the <u>program</u> to compute $f_k := f(\tau_k)$, assuming τ_k in X and putting f_k into X. R_1 is available for additional storage. The instruction $\bar{x}$ (see 05 and 17) has been used in order to save programming space; together with $\frac{k}{n} = \tau_k$ it furnishes another quotient (in Y), which is not needed.

5. <u>Operating Instructions</u>

Load program, including program to compute f_k. Move to RUN. Select mode of displaying numbers. Load data as follows:

1 into R_2 and R_3
0 into R_4, R_5, R_6, R_7 ,

and τ, the value at which the polynomials t_ℓ should be evaluated, into the X-register. Press

PRGM
R/S

At the first stop $t_0(\tau)$ is shown. Press

R/S

to obtain $t_1(\tau)$, etc.

6. Examples and Timing

1. $f(\tau) := |2(\tau - [\tau]) - 1|$, see Essentials, Demonstration 7.2-3, when $[\tau]$ denotes the largest integer $\leqq \tau$. For $\tau > 0$ the program computing f_k can be written as:

06	ENTER	or	06	FRAC
07	INT		07	2
08	-		08	*
09	2		09	1
10	*		10	-
11	1		11	ABS
12	-		12	NOP
13	ABS		13	NOP
14	NOP		14	NOP
15	NOP		15	NOP

(If $\tau < 0$ the INT operation yields $[\tau] + 1$, but we know that $f(\tau) = f(-\tau)$ for any τ, see Figure 7.2b.) The computing time for $\ell = 1$ is approx. 4 sec. and doubles at each step; for $\ell = 14$ it is about 6 hrs.

2. $f(\tau) := (1 + \varepsilon \cos 2\pi\tau)^{-1/2}$, see Essentials, Demonstration 7.4-1, where $\varepsilon = 0.9$ and $\tau = 0.2$. If we store the constant 2π in R_1, the program computing f_k is:

06	RCL 1	11	*
07	*	12	1
08	cos	13	+
09	.	14	$\sqrt{x}$
10	9	15	1/x

Note: If the program given in section 4 above should be used for a function, which to program takes a little more than 10 instructions, then the main program can slightly be shortened e.g. by

- storing τ into R_0 "by hand"
- pressing RAD "outside"
- rearranging the program such that $t_\ell(\tau)$ is in the X-register after instruction 49 has been completed (and thereby saving one instruction R/S)

Demonstrations 7.3-1, 7.3-2, 7.3-3:

COMPLEX FOURIER COEFFICIENTS, DATA GENERATED, PERIOD 1

1. Purpose

To compute the complex Fourier coefficients of a given function f with period 1.

2. Method

For $k := k_0, k_0+1, \ldots$ we compute the trapezoidal values $\hat{c}_k$ of the complex Fourier coefficients c_k by

$$\hat{c}_k = \frac{1}{n} \sum_{m=0}^{n-1} f\left(\frac{m}{n}\right) e^{-\frac{2\pi i k m}{n}},$$

see Essentials, §7.3 (also for estimates of the error $\hat{c}_k - c_k$, e.g. Theorem 7.3b). We assume that f is given by a formula and hence have to generate the values f_k of f at the sampling points $\tau_k := k/n$. Letting

$$\theta := \frac{2\pi k}{n},$$

since exp (any integer multiple of $2\pi i$) = 1, the foregoing formula can be written as

$$\hat{c}_k = \frac{1}{n} \sum_{m=1}^{n} f_{n-m} \, e^{-im\theta} \; .$$

3. <u>Flow Diagram</u>

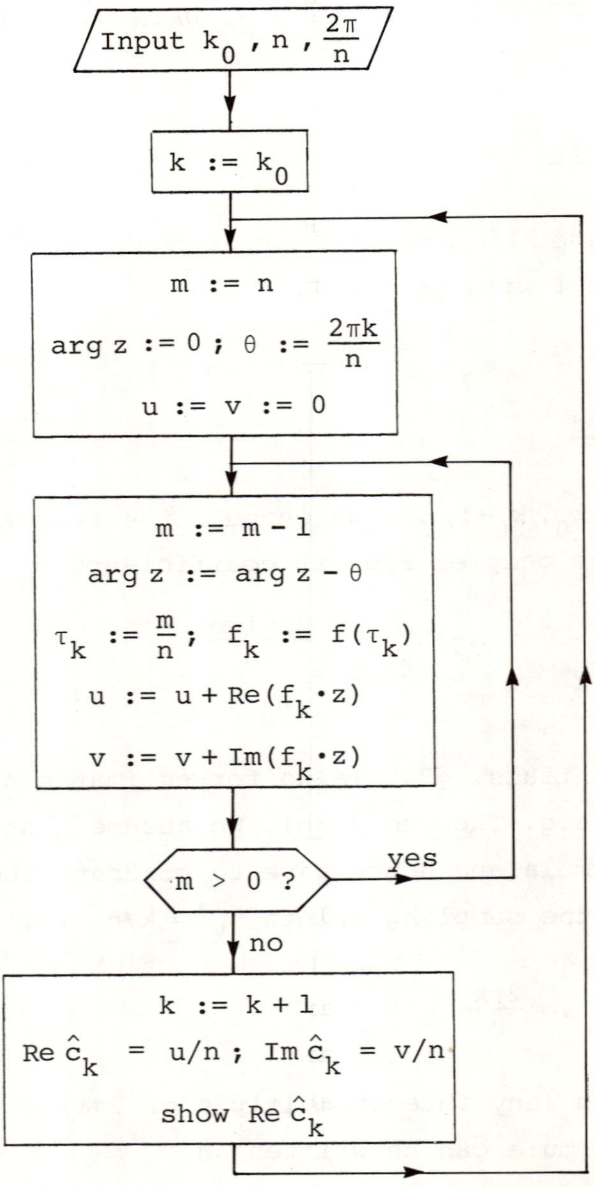

Since $|z| := |e^{im\theta}| = 1$, using polar coordinates, in order to multiply $f_k \cdot z$, only the arguments of f_k and z have to be added and $|f_k z| = |f_k|$; therefore $|z|$ is not explicitly used. See also the remarks to the flow diagram of "Slow Fourier Transform".

4. Storage and Program

R_0	R_1	R_2	R_3	R_4	R_5	R_6	R_7
	$\boxed{k_0}$	$\boxed{n}$	m	$\boxed{\frac{2\pi}{n}}$	θ	u	v

	00	10	20	30	40
0	X	1	STO 6		
1	RCL 2	STO-3	R ↓		RCL 3
2	STO 3	RCL 7	STO 7		x > 0
3	RCL 1	RCL 6	$\overline{x}$		GTO 10
4	RCL 4	→ P			1
5	*	x ≷ y			STO+1
6	STO 5	RCL 5			RCL 2
7	0	-			STO÷6
8	STO 6	x ≷ y			STO÷7
9	STO 7	→ R			RCL 6

Instructions 24 ÷ 40 are reserved for the program computing f_k. This program should assume τ_k in X and add the real part of f_k to the content of R_6, and add the imaginary part of f_k to the content of R_7. R_0 is available for additional storage.

5. Operating Instructions

Load program, including program to compute f_k. Move operating switch to RUN. Press

RAD

(otherwise the results will be wrong!). Select mode of displaying numbers and load data as follows:

k_0	into	R_1
n	into	R_2
$\frac{2\pi}{n}$	into	R_4

The computation is started by pressing

PRGM
R/S

The calculator will stop, displaying the real part of the first computed Fourier coefficient $\hat{c}_k$ $(k = k_0)$. Press

RCL 7

to obtain the imaginary part of $\hat{c}_k$. To compute $\hat{c}_{k+1}$ press

R/S

At stop $\mathrm{Re}\,\hat{c}_{k+1}$ will be displayed and $\mathrm{Im}\,\hat{c}_{k+1}$ stored in R_7, etc.

6. Examples

[1] $f(\tau) := |2\tau - 1|$, see Essentials, Demonstration 7.3-1. Program to compute and store f (i.e. here, since f is real-valued, add it to the contents of R_7, see section 4):

```
24   2          30   NOP
25   *            ...
26   1          40   NOP
27   -
28   ABS
29   STO+6
```

For $n = 4,8$ it is interesting to compute all the coefficients $\hat{c}_k$, $k = 0, 1, \ldots, 7$; thus it can be verified, that the coefficients with even index are zero (within the accuracy used on the calculator) and that $\hat{c}_k = \hat{c}_{k+n}$. For $n \geqq 16$ the computations are more time-consuming and we therefore only compute the coefficients with odd indices. To this end we let $k_0 := 1$ and change instruction 44 to "2" (this corresponds to $k := k + \underline{2}$ in the last box of the flow diagram in section 3). To verify that the $\hat{c}_k$ are real, press

RCL 6

to obtain $\text{Im}\,\hat{c}_k = 0$. The accelerated values (for $k = 1$) have been computed "by hand".

[2] $f(\tau) := (1+\varepsilon\cos 2\pi\tau)^{-1/2}$, $|\varepsilon| < 1$, see Essentials, Demonstration 7.3-2. f again is real-valued. If ε is stored in R_0, the program to compute and store f can be written as:

24	π	30	*	36	NOP
25	*	31	1	37	NOP
26	2	32	+	38	NOP
27	*	33	$\sqrt{x}$	39	NOP
28	cos	34	1/x	40	NOP
29	RCL 0	35	STO+6		

After loading the program (with instruction 44 changed back to "1"), $\varepsilon = 0.9$ has to be stored into R_0.

[2] $f(\tau) := e^{ix\cos 2\pi\tau}$, x parameter, see Essentials, Demonstration 7.3-3. Program to compute and store f, assuming x in R_0:

24	2	30	*	36	NOP
25	*	31	1	37	NOP
26	π	32	$\rightarrow$ R	38	NOP
27	*	33	STO+6	39	NOP
28	cos	34	$x \gtrless y$	40	NOP
29	RCL 0	35	STO+7		

The foregoing program computes $f(\tau)$ in polar coordinates first (instr. 24 ÷ 30: argument, instr. 31: modulus) and then transforms it to rectangular coordinates. Before starting the computation $x = 4.5$ has to be loaded into R_0. It is seen that

$$\hat{c}_k = \begin{cases} \text{real,} & \text{k even} \\ \text{purely imaginary,} & \text{k odd} \end{cases}$$

The program given in section 4 automatically only displays $\operatorname{Re}\hat{c}_k$. To obtain $\operatorname{Im}\hat{c}_k$ press

RCL 7

Demonstrations 7.6-1, 7.7-6, 7.7-7, 7.8-1:

FAST FOURIER TRANSFORM

1. Purpose

Given a data vector $\underline{x} = (x_k)$ of length $n = 2^\ell$, $k = 0, 1, 2, \ldots, n-1$. To compute the Fast Fourier Transform $\underline{y} = \underline{F}_n \underline{x}$ (or its conjugate $\overline{\underline{F}}_n \underline{x}$), abbreviated FFT, of $\underline{x}$, see Essentials, §7.5.

2. Method

The basic idea in the construction of the FFT is to use the duplication theorem 7.5a of Essentials repeatedly. The program given in section 4 below, is based on the implementation of the FFT, as described in Essentials, §7.6, where the storage requirements have been kept low. We recall the crucial definitions and formulas. The bit reversion function p(m) assigns to the integer $m < 2^j$ with binary representation $m_0 m_1 \ldots m_{j-1}$ the integer m' with binary representation $m_{j-1} m_{j-2} \ldots m_0$. The 2^j vectors of length $2^{\ell-j}$, $j = \ell, \ell-1, \ldots, 0$ in each row of the scheme in Fig. 7.6d are combined into a single vector $\underline{y}_j := (y_j[k])$ (from now on we use brackets instead of indices for the components of a vector); the vector $\underline{y}_0$ in the top row of the scheme is the FFT of $\underline{x}$, which we have to construct, starting from the bottom row $\underline{y}_\ell$. Theorem 7.6a yields

$$y_\ell[m] = x[p_\ell(m)] ,$$

$m = 0, 1, \ldots, 2^\ell - 1$. In other words: The bit reversion function has to be applied to the given data vector first, and the resulting vector equals $\underline{y}_\ell$. The duplication theorem 7.5a yields for the upper rows ($j \leq \ell-1$), using the abbreviations

$$y := y_j , \quad y' := y_{j-1} \quad (j = \ell, \ell-1, \ldots, 1) ,$$

$$s := 2^{\ell-j} ,$$

$$r_j := e^{2\pi i \cdot 2^{-j}} , \quad r_{\ell-j+1} =: e^{i\phi} , \quad r^h_{\ell-j+1} =: e^{i\psi} ,$$

$$y'[h] = \frac{1}{2}\left\{y[h] + e^{i\psi} y'[h+s]\right\} ,$$

$$y'[h+s] = \frac{1}{2}\left\{y[h] - e^{i\psi} y'[h+s]\right\} ,$$

where h, defined as $h := 2 \cdot s \cdot m + k$, where $m = 0, 1, \ldots, 2^j - 1$ and $k = 0, 1, \ldots, s-1$, runs from 0 to s.

3. Flow Diagram

See Essentials, Fig. 7.6e. In the program for the HP-33E calculator for reasons of programming space the bit inversion function has to be applied by hand, and for reasons of storage space the input of the data has to be done gradually (see section 5). The vector $\underline{y}'$ is computed by the formulas given in section 2. In terms of the ("little") vectors in the scheme of Fig. 7.6d the three loops in the flow diagram can be explained as follows:

If $|\psi| < \pi$ (program tests $h + s < 2^j$) then compute corresponding element of other vector of pair of vectors ($j + 1$ pairs).

If $h < n$ then compute next pair of vectors.

If $s < n$ then compute next (upper) row.

4. Storage and Program

R_0	R_1	R_2	R_3	R_4	R_5	R_6	R_7
Re y	Im y	[n]	k=h+s	h	[1]	arg p	[±π]
					2^j		ϕ

[RAD]

	00	10	20	30	40
0	╳	+	STO−1	R/S	RCL 5
1	0	R/S	2	RCL 7	STO+4
2	STO 4	→ P	STO÷0	STO+6	RCL 4
3	0	x ≷ y	STO÷1	1	RCL 2
4	STO 6	RCL 6	CLX	STO+3	x > y
5	STO 3	+	RCL 1	STO+4	GTO 03
6	RCL 4	x ≷ y	+	RCL 3	2
7	R/S	→ R	x ≷ y	RCL 5	STO*5
8	RCL 4	STO−0	RCL 0	x > y	STO÷7
9	RCL 5	x ≷ y	+	GTO 06	GTO 01

As in previous programs, the option of transforming into polar (→ P) and back to rectangular (→ R) coordinates has been used to implement the arithmetic with complex numbers.

5. Operating Instructions

Load program, move to RUN, select mode of displaying numbers. Load data as follows:

n	into	R_2	
1	into	R_5	
$-\pi$ / $+\pi$	into	R_7	if $\underline{F}_n\underline{x}$ / $\overline{\underline{F}}_n\underline{x}$ is to be computed.

Apply the bit inversion function (see section 2) manually to $\underline{x}$. Hence at the beginning the position h contains $y_{p[h]}$. Thus for $n=8$ the starting situation is

h	$y_{p[h]}$
0	y_0
1	y_4
2	y_2
3	y_6
4	y_1
5	y_5
6	y_3
7	y_7

The typical operation is

$$y'_h = \frac{1}{2}[y_h + py_k]$$
$$y'_k = \frac{1}{2}[y_h - py_k] \qquad (*)$$

where $y = \mathrm{Re}\, y + i\, \mathrm{Im}\, y$ are complex numbers, h and k indices, $0 \leq h, k \leq n$, and p is a certain root of unity. The computation of each new pair y'_h, y'_k according to (*) involves 3 R/S stops. At the <u>first</u> stop, the index h is shown. The operator is to load

$\mathrm{Re}\, y_h$ into R_0, $\mathrm{Im}\, y_h$ into R_1,

At the <u>next</u> stop, the index k is shown. The operator is to load

$\mathrm{Re}\, y_k$ into X, $\mathrm{Im}\, y_k$ into Y.

(Thus $\mathrm{Im}\, y_k$ should be loaded first.) At the last stop

$\mathrm{Re}\, y'_h$ is in X, $\mathrm{Im}\, y'_h$ is in Y

$\mathrm{Re}\, y'_k$ is in R_0, $\mathrm{Im}\, y'_k$ is in R_1

These quantities should be recorded in the h and k positions of the new column. Now (do not forget to) press

RAD

and

PRGM

R/S

to start the computation.

6. Examples

[1] Let $\underline{x} = (x_k)$, where $x_k := f(\frac{k}{8})$, $k = 0, 1, \ldots, 7$, $f(\tau) := \exp(ix \cos 2\pi\tau)$ and the parameter x set to 4.5. To compute $\underline{F}_8\underline{x}$, see Essentials, Demonstration 7.6-1. This example is identical with Demonstration 7.3-3, where the slow Fourier transform of $\underline{x}$ was computed. It can be verified for $n = 8$ that the slow and the fast FT yield the same results.

[2] To smooth the periodic sequence

$\underline{x} := \|: 7\ 4\ 8\ 3\ 2\ 9\ 6\ 5 :\|$

by the smoothing sequence

$\underline{s} := \frac{1}{4} \|: 2\ 1\ 0\ 0\ 0\ 0\ 0\ 1 :\|$,

as explained in Essentials, Example 7.7-3, see Essentials, Demonstration 7.7-6. According to the Convolution Theorem 7.7a we first compute $\hat{\underline{x}} := \underline{F}_8\underline{x}$ (using the program of section 4 with $\phi = -\pi$) and $\hat{\underline{s}} := \underline{F}_8\underline{s}$ (can be done analytically). Then we manually perform the multiplications $n^2 \cdot \hat{s}_k \cdot \hat{x}_k$, $n = 8$, $k = 0, 1, \ldots, 7$. Finally $\overline{\underline{F}}_8(n^2 \cdot \hat{s} \cdot \hat{x})$ (using the program of section 4 with $\phi = +\pi$) is computed.

3 To compute the product 241 * 769 by the convolution theorem see Essentials, Demonstration 7.7-7 and Example 7.7-5. With

$$\underline{p} := \| : 1, 4, 2, 0, 0, 0, 0, 0 : \|$$

$$\underline{q} := \| : 9, 6, 7, 0, 0, 0, 0, 0 : \| ,$$

$n = 8$, we have to compute $\hat{\underline{p}}, \hat{\underline{q}}$ (program of section 4 with $\phi = -\pi$), then $64 \cdot \hat{\underline{p}}\hat{\underline{q}}$ (by hand) and finally $\overline{\underline{F}}_8(64 \cdot \hat{p}\hat{q})$ (program of section 4 with $\phi = +\pi$).

4 Let $\{0.2, 0.9, -0.4, -0.1\}$, $m = 4$, be a time series. To compute the covariance function $\underline{r}$, the power spectrum $\underline{f}$ and the smoothed power spectrum $\underline{g}$ (weights $\frac{1}{4}, \frac{1}{2}, \frac{1}{4}$) (see Essentials, §7.8 for definitions), see Essentials, Demonstration 7.8-1. Two methods are available:

Method A (conventional)

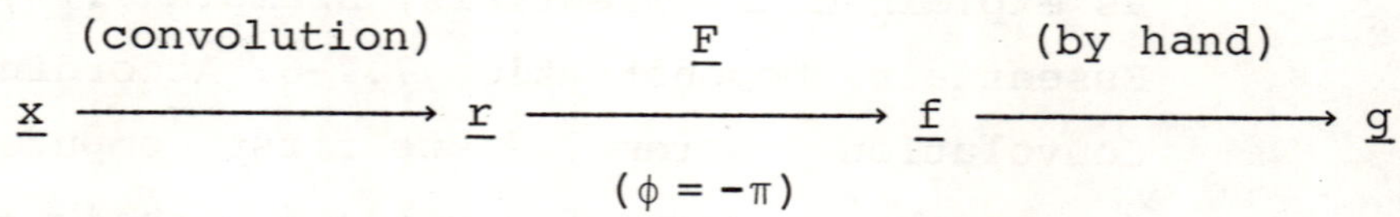

Method B (avoids convolutions)

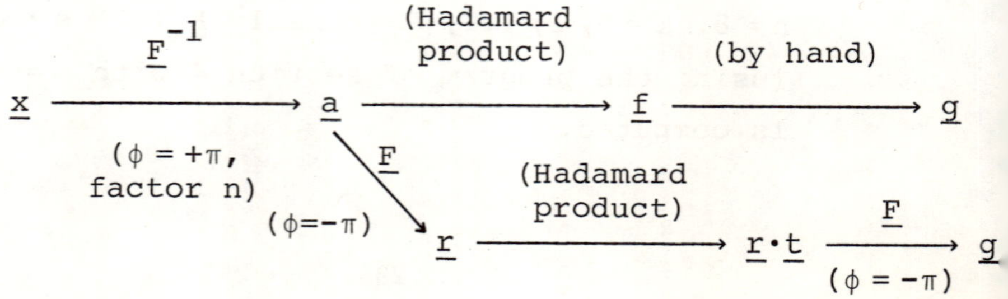